Introduction to
Environmental Virology

Introduction to Environmental Virology

GABRIEL BITTON

Department of Environmental Engineering Sciences
University of Florida
Gainesville, Florida

A WILEY-INTERSCIENCE PUBLICATION

JOHN WILEY & SONS
New York • Chichester • Brisbane • Toronto

Library of Congress Cataloging in Publication Data:

Bitton, Gabriel.
    Introduction to environmental virology.

    "A Wiley-Interscience publication."
    Includes index.
    1. Sanitary microbiology.   2. Virus diseases—
Prevention.   3. Waterborne infection—Prevention.
4. Indicators (Biology)   I. Title.   [DNLM:
1. Environment.   2. Viruses.   QW160 B624i]

QR48.B5   576'.165   80-13469
ISBN 0-471-04247-1

Printed in the United States of America

10 9 8 7 6 5 4 3 2 1

Humbly dedicated to the memory of my mother and to Nancy, Julie, and Natalie

A book must be the axe
for the frozen sea inside us.

Franz Kafka

Preface

Environmental virology is now a discipline in its own right, and this book attempts to strengthen its establishment and encourage its further development.

The mission of environmental virology is to provide information about the behavior of pathogenic viruses outside their hosts, their contribution to disease occurrence in man, and means to improve their control by the numerous barriers that have been established through modern technology. The last two decades have witnessed a significant change in outlook and methodology in the field of environmental virology, which has attracted scientists from various disciplines including virology, microbiology, hydrology, soil science, geology, engineering, and marine biology. The next decade will witness an emphasis on virus monitoring, epidemiology of waterborne viral diseases, and a better assessment of viruses as an environmental issue.

It is my hope that this book will help bridge the gap between microbiologists, particularly virologists, and environmental engineers and scientists. Throughout this book, particularly in chapters dealing with water and wastewater treatment, I have attempted to provide the reader with both the engineering and the virological aspects of the subject. To put the virus aspect into perspective I have also provided information concerning the fate of bacterial pathogens and protozoan and helminth parasites.

This book is designed as a textbook in environmental virology or as supplementary material in environmental microbiology courses. It should also be useful as a reference book for environmental engineers and scientists, agricultural engineers, water and wastewater treatment experts, food scientists, and public health workers.

I am most grateful to many colleagues, friends, companies and publishers who have provided illustrations for this book. I sincerely appreciate the useful and inspiring discussions held with colleagues during scientific meetings around the world. Special thanks are extended to Drs. Charles Gerba and Samuel Farrah for providing me with preprints as soon as they became available. I acknowledge the hard work and dedication of my graduate students, who have contributed a great deal to my understanding of environmental virology.

I am most grateful to Dean Bernard Sagik and Dr. Charles Gerba who have provided me with helpful and constructive criticism.

My efforts were greatly aided by the dedicated labor of Suellen Patton, who typed the manuscript, and Daniel Tillman and Wesley Bolch, who made the drawings for most of the figures. My deep gratitude is extended to Dr. Mary

Conway of Wiley-Interscience for her interest and encouragement during the preparation of this book.

I am most grateful to my wife, Nancy, who typed part of the manuscript despite her many responsibilities. This book would not have been possible without her patience, understanding, and moral support.

GABRIEL BITTON

Gainesville, Florida
March 1980

Contents

one
Viruses:
General Properties and Classification

1.1 GENERAL PROPERTIES OF VIRUSES

1.1.1 Size

Viruses are the smallest obligate intracellular parasites that infect and some-times cause a variety of diseases in animals, plants, bacteria, fungi, or algae.

1

The size range of living organisms is shown in Figure 1.1. It is clear that viruses are colloidal particles ranging from 20 nm (200 angstroms) to approximately 350 nm (3500 angstroms).

The observation of the size and shape of viruses was made possible with the advent in 1931 of the electron microscope, which allows the easy observation of virus particles such as those of poliovirus, which measure approximately 25 nm (Figure 1.2). In the early days of microbiology viruses were designated as "filterable agents" because they are able to pass through conventional filters normally used for the retention of bacterial cells.

1.1.2 Anatomy of a Virus

With the advent of electron microscopy and x-ray diffraction it became possible to obtain more detailed information about the various components of a virus

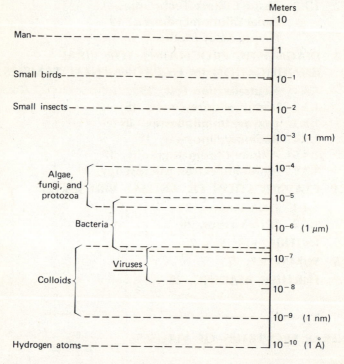

FIG. 1.1. A schematic representation of the size range (logarithmic scale) of living organisms and of colloidal particles. From K. C. Marshall (1976), *Interfaces in Microbial Ecology,* Harvard University Press, Cambridge, Mass. Courtesy of Harvard University Press.

FIG. 1.2.   Dispersed preparation of poliovirus (Mahoney strain), which has been purified by Freon extraction and sucrose gradient centrifugation. Courtesy of R. Floyd.

particle. A *virion* is a mature infectious particle composed of a *core* of *nucleic acid* (RNA or DNA) surrounded by a protein coat called *capsid*. The capsid is made of an arrangement of a variable number of subunits known as *capsomeres*. The combination of capsid and nucleic acid core is called *nucleocapsid*. In some viruses the nucleocapsid is surrounded by an *envelope* that contains lipoproteins or lipids. Viruses devoid of such an envelope are called *naked virions* (Figure 1.3).

The protein coat confers specificity on the virion and protects it from adverse environmental conditions (see Chapters 3 and 4). The nucleic acid core confers infectivity to the virion. It may contain four possible types of nucleic acid: single- or double-stranded DNA and single- or double-stranded RNA. The most common types of nucleic acids found in viruses are single-stranded RNA and double-stranded DNA.

The capsid architecture is an important feature to consider in the characterization of viruses. One generally recognizes two main classes in capsid symmetry.

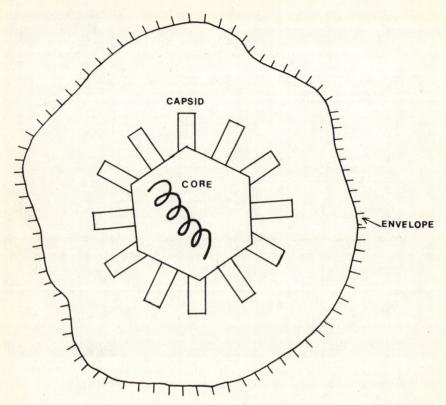

FIG. 1.3.   Virus structure. Adapted from R. W. Horne (1963), *Sci. Am.* **208:**48–56.

- *Helical symmetry.* The capsid is a cylinder with a helical structure. The single-stranded RNA of tobacco mosaic virus (TMV) is packed in a helically arranged capsid.
- *Polyhedral symmetry.* The capsid is an *icosahedron* which has 20 triangular faces, 12 corners, and 30 edges (Figure 1.4). This type of symmetry is found in animal viruses such as poliovirus or adenovirus.

The shape, size, and nucleic acid type of some animal, plant, and bacterial viruses are shown in Table 1.1.

1.1.3   General Classification of Viruses

Various systems have been proposed for the classification of viruses. The following features are generally considered:

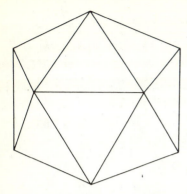

FIG. 1.4.   Polyhedral symmetry: Icosahedron.

1. *Type of Nucleic Acid.* Unlike bacteria, viruses contain either RNA or DNA genomes. Furthermore, the nucleic acid may contain one or two strands.

2. *Number of Capsomeres in the Capsid.* For example, the enterovirus capsid contains 32, whereas adenoviruses contain 252 capsomeres.

3. *Virus Size.*

4. *Capsid Architecture.* Polyhedral or helical structure.

5. *Presence of an Envelope Around the Capsid.*

6. *Susceptibility of the Virion to Chemical Agents.* (For example, ether or chloroform.) Viruses containing lipids in their envelope (e.g., influenza virus) are inactivated by these chemical agents.

7. *Site of Replication of the Nucleic Acid.* This process may occur in the cytoplasm or in the nucleus of the host cell.

8. *Host Cell for the Virus.* Viruses are now known to parasitize animal cells, bacteria, plant cells, blue-green algae, and fungi. According to the infected host cell the following can be distinguished:

    (a) *Animal Viruses.* (Vertebrates and invertebrates.) Their classification and general features are examined in Section 1.4.

    (b) *Bacterial Phages or Bacteriophages.* They are known to infect a wide range of bacteria. A typical phage is made of a *head* (capsid), which houses the nucleic acid core; a *sheath* or "tail," which is attached to the head through a "neck"; and *tail fibers,* which help in the adsorption of the virion to the host cell (Figure 1.5). The capsid contains mostly double-stranded DNA, but one may find phages with a single-stranded DNA (ϕX 174) or RNA (f2 and MS2 phages).

TABLE 1.1. Shape, Size, and Nucleic Acid Type of Some Viruses[a]

	Virus	Nucleic Acid Type	Shape	Size (nm) (or diameter × length)
Animal Viruses	Vaccinia	DNA		230 × 300
	Mumps	RNA		150 × 300
	Herpes	DNA		100 × 200
	Influenza	RNA		80 × 120
	Adenovirus	DNA		70 × 90
	Poliovirus	RNA		28
Plant Viruses	Wound tumor	RNA		55 × 60
	Tobacco mosaic	RNA		18 × 300
	Potato X	RNA		10 × 500
Bacterial Phages	T phage	DNA		65 × 200
	φX174	DNA		25

[a] Adapted from H. Lechevalier and D. Pramer (1971), *The Microbes*. J. B. Lippincott Co., Philadelphia, Pa., and R. W. Horne (1963), "The Structure of Viruses," *Scientific American*.

It has been proposed to group bacteriophages in six categories based on type of nucleic acid, presence or absence of a tail, capsid architecture, and contractibility of the tail (Table 1.2).

Bacteriophages were discovered in human feces by d'Herelle in 1917. Coliphages (i.e., phages infecting *Escherichia coli*) are widespread in wastewater and other bodies of water that have been contaminated with sewage. Since their isolation and assay are easy and inexpensive, it has been proposed to use them as indicators of viral pollution. This topic is discussed in more detail in Chapter 12. Meanwhile, numerous researchers in the field of

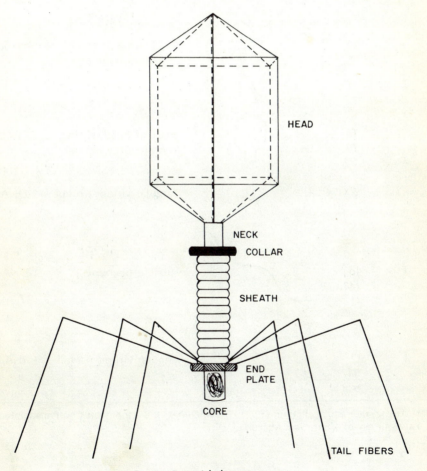

FIG. 1.5. Bacterial phage structure.

TABLE 1.2. Major Properties of the Basic Morphologic Types of Bacteriophages[a]

Bradley Type	Examples	Basic Morphology	Description	Nucleic Acid Type
A	T-even		Angular head, contractile tail	2-DNA
B	T1 T5		Angular head, long noncontractile trail	2-DNA
C	T3 T7		Angular head, short noncontractile tail	2-DNA
D	φX174		Angular virion, no tail	1-DNA
F	fd M13 f1		Flexible filament	1-DNA
E	f2 R17 fr		Angular virion, no tail	1-RNA

[a] H. Lechevalier and D. Pramer (1971), *The Microbes*, J. B. Lippincott Co., Philadelphia, Pa. Courtesy of Amer. Soc. Microbiol.

environmental virology have used bacteriophages as model viruses and have sometimes found a similarity between these viruses and those of animal origin. Phages have been used to evaluate the efficiency of water and wastewater treatment, elucidate viral inactivation mechanisms, and improve virus detection methodology.

(c) *Blue-green Algal Viruses.* This group of viruses is relatively new to the field of virology, having been discovered in 1963 by Safferman and Morris. They attack a number of blue-green algae and are usually called *cyanophages.* These phages are generally named after their algal host; for example, LPP_1 cyanophage has a series of three algal hosts: *Lyngbia, Phormidium,* and *Plectonema.* Their host range is rather narrow in general. Morphologically these resemble the bacteriophages described in 8(b) above, that is, a tail attached to a polyhedral head. Their size ranges from 20 to 250 nm, and they all contain DNA. Table 1.3 shows some properties of the major groups of cyanophages.

Algal viruses are ubiquitous in nature. They have been isolated around the world from oxidation ponds, rivers, or fish ponds. Blue-green algae are known to undergo cyclic blooms that are responsible for massive fishkills in bodies of water. The causes of these blooms are only partially understood. It has been suggested that cyanophages may control the distribution and population dynamics of blue-green algae. However, other factors are probably involved in this control. Some have claimed that

TABLE 1.3.   Some Characteristics of Major Groups of Cyanophages[a]

Code Name of Cyanophage	LPP_1 LPP_2	SM-1	N-1	AS-1
Algal host(s) (genera)	Lyngbia Phormidium Plectonema	Synechococcus Microcystis	Nostoc	Anacystis Synechococcus
Nucleic acid	DNA	DNA	DNA	DNA
Size (length, width in nm)	20–15	—	110–16	243.5–22.5
Head	Polyhedron (six-sided outline)	Icosahedron	Polyhedron (six-sided outline)	Polyhedron (six-sided outline)

[a] Adapted from E. Padan and M. Shilo (1973), *Bacteriol. Rev.* **37**:343–370.

these viruses can be used for the biological control of blue-green algae, but this still remains at the experimental stage.

(d) *Fungal Viruses*. Fungal viruses have been reported for each of the major taxonomic groups of fungi. However, only a few of them have been studied in detail. The most extensively studied fungal virus has been the one that attacks *Penicillium chryso-genum*, the mold used for the commercial production of penicillin.

Fungal viruses are relatively small; their size ranges from 33 to 41 nm. They all contain double-stranded RNA. Few of them produce lysis of their host cell, and apparently healthy mycelia may contain high numbers of virus particles. Since pathogenic fungi are important parasites of higher plants, it has been suggested that fungal viruses may be useful in the biological control of fungal diseases in plants.

1.1.4   Reproduction of Viruses

The mechanisms involved in the lytic cycle of viruses have been elucidated by using bacteriophages as model viruses. Basically a similar sequence of events is found in both bacterial and animal viruses:

1. *Adsorption*. This is the first step in the reproduction cycle of viruses. The virus particle must adsorb to *receptor sites* on the host cell. This step has been well studied in the case of the T series of bacteriophages that adsorb to the host cell by their tail fibers. Animal viruses adsorb to specific surface components of the host cell. The receptor sites involved are varied. For myxoviruses the receptor is a mucoprotein, while the receptor site for poliovirus is a lipoprotein. Little is known, however, about the molecular reactions involved in the adsorption process. Nevertheless, for both types of viruses the attraction and adsorption to the host cell is probably of electrostatic nature and is sometimes dependent on salt concentration and pH of the suspending medium.

2. *Penetration*. This step varies with the virus under consideration. Bacteriophages "inject" their nucleic material into the host cell. The capsid remains outside the cell and is called a *ghost*. For animal viruses the whole virion penetrates the host cell wall. It has been advanced that the host cell engulfs the virion, and this process is called *phagocytosis*. However, the passage of some viruses through the host cell membrane is still poorly understood.

3. *Eclipse*. Once inside the host cell, animal viruses are "uncoated" (stripping of the protein coat), and this results in the liberation of nucleic

acid. This stripping is done through the action of the host proteolytic enzymes.

4. *Replication.* The replication site within the host cell varies with viruses. It can be either the cytoplasm or the nucleus. This step results in the replication of the virus nucleic acid and in the synthesis of viral proteins.

5. *Maturation.* This involves the assembly of nucleic acid and protein coat to form a nucleocapsid.

6. *Release of Mature Virions.* Virus release is generally due to the rupture of the host membrane but may result from "budding off" of the membrane.

1.2 TECHNIQUES FOR VIRUS CULTIVATION AND ENUMERATION

1.2.1   Animal Inoculation

Inoculation was a traditional method for detecting animal viruses prior to the discovery of *in vitro* culture on animal cells. Experimental animals (mice) are infected with the suspected infectious agent by using various routes of inoculation (intravenous, intracerebral, intranasal, or intraperitoneal). With the advent of tissue culture techniques this method has become less popular, although it is essential for the isolation of certain coxsackie A viruses (enteric viruses).

1.2.2   Chicken Embryo Inoculation

Chicken eggs are currently used for the isolation of certain viruses like influenza virus. The allantoic and amniotic cavities within the egg are routinely used as injection sites. For certain viruses (e.g., poxviruses) one generally counts the lesions or *pocks* that appear on the surface of the chorioallantoic membrane. Chicken embryos cannot, however, be used for the isolation of enteric viruses.

1.2.3   Tissue Culture Techniques

Tissue culture technology opened a new era for the cultivation of animal viruses under laboratory conditions. Viruses were first propagated only with intact animals, and the first attempt to grow them in tissue culture dates back to 1913. In 1949 there was a major breakthrough in tissue culture technology: John Enders and his collaborators reported the successful growth of poliovirus (a neurotropic agent) in nonneural tissue culture and received the Nobel Prize in 1954 for their efforts. Refinements in aseptic techniques and tissue culture media and the use of antibiotics to control bacterial and fungal contamination led to further advances in tissue culture technology.

There are two types of cell lines:

- *Primary Cell Line.* These cells are removed directly from the host tissues (e.g., kidney) and can be subcultured for only a limited number of times.

- *Continuous Cell Line.* Animal cells, after serial subculturing, acquire characteristics that are different from the original cell line, allowing them to be subcultured indefinitely.

These cell lines can be subcultured conveniently in balanced salt media supplemented with serum. Their culturing must be undertaken under aseptic conditions to avoid bacterial and fungal contamination. The addition of antibiotics to the growth media generally helps in avoiding some of these problems.

Under appropriate nutritional conditions animal cells grow and form a *monolayer* on the inner surface of a glass or plastic bottle. (Neoplastic cells behave differently and form a multilayer of cells called a *tumor.*)

We will now examine how infective viral particles can be determined, using tissue culture technology.

Serial Dilution Endpoint

Aliquots of serial dilutions of a viral suspension are inoculated into cultured host cells (three to five tubes) or into susceptible animals. After appropriate incubation changes that have occurred as a result of virus replication are measured. With the aid of a microscope morphological changes [cytopathic effect (CPE)] resulting from the growth of viruses can be observed. Death, paralysis, or some other change in intact animals (e.g., mice) is recorded. The percentage of positive samples at each serial dilution is recorded and plotted against viral dilution (Figure 1.6). The titer or *endpoint* is the highest viral dilution (i.e., smallest amount of viruses) capable of producing CPE in 50% of the cultures and is referred to as $TCID_{50}$ (tissue culture infectious dose). If animals are used ID_{50} (infectious dose that produces a change in 50% of the animals) or LD_{50} (lethal dose that causes death in 50% of the animals) are the terms used to express the viral titer.

Plaque Assay Method

An aliquot of viral suspension is placed on the surface of a cell monolayer. After an appropriate incubation period to allow maximum adsorption of the virus to the host cells an overlay of soft agar or carboxymethylcellulose is

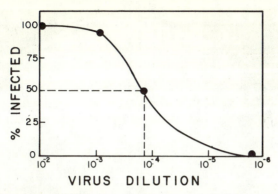

FIG. 1.6. Determination of TCID$_{50}$. Adapted from R. F. Boyd and B. G. Hoerl (1977), *Basic Medical Microbiology*, Little, Brown and Co., Boston, Mass.

FIG. 1.7. (A) Plaque of poliovirus type 1 (Sabin strain). The virus was assayed on human amnion cells (AV3 line). (B) Plaques of echovirus 1 (assayed on simian host cell, MA104).

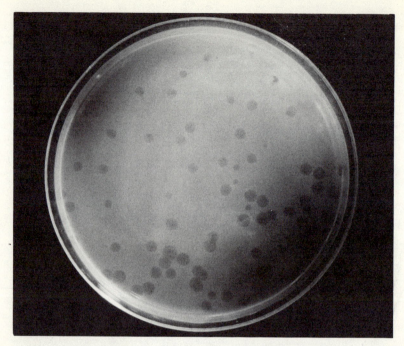

FIG. 1.8.   Plaques of bacteriophage T7.

poured on the surface of the monolayer. Each initially infected cell leads to a *zone of infection* where the cells have been destroyed. This localized area of cell destruction is called a *plaque*. Vital stains incorporated in the overlay help in the differentiation between dead and live host cells. The plaque technique offers a precise way of quantifying viruses and can detect more than one virus type on the same monolayer. Unfortunately, it is expensive and requires well-trained personnel.

The results are expressed in numbers of *plaque-forming units* (PFU), which appear on the cell monolayer. Poliovirus and echovirus plaques are illustrated in Figure 1.7.

Bacteriophages are also assayed by a plaque assay method based on similar principles. They form plaques that are zones of lysis of the bacterial lawn (Figure 1.8).

1.3   DIAGNOSTIC PROCEDURES FOR VIRAL IDENTIFICATION IN ENVIRONMENTAL SAMPLES

Although diagnostic procedures for viral identification are costly and time-consuming, they are sometimes necessary for the identification of viral agents involved in serious epidemics. The use of tissue cultures in the diagnosis of viral agents does not give any indication of their virulence. This is known only through examination of genetic markers such as the *d marker* or ability to grow under an acid overlay and the *T marker* or ability to grow at 40.5°C in poliovirus. This examination is limited only to viruses like the three strains of poliovirus. Tissue culture techniques are further limited by the following:

- They are not suitable for the growth of certain viruses (e.g., certain coxsackievirus A serotypes).
- The growth of certain viruses is masked by faster growing ones.
- Host cell toxicity occurs with some environmental samples.

In addition to the traditional serological tests (neutralization test, immunofluorescence, and hemagglutination inhibition test) new ones have been developed, and they contribute greatly to the field of diagnostic virology.

1.3.1 Neutralization Test

The neutralization test is based on the neutralization of virus by specific antibodies found in the serum. After incubation for 2 hours at 37°C the serum–virus mixture is inoculated into tissue cultures, which are examined for CPE after several days of incubation.

Under the sponsorship of the National Institute of Health 42 antisera have been combined into 8 pools called the *Lim Benyesh–Melnick pools* (LBM). Each pool contains 10 to 11 antisera. With the LBM kit up to 37 enteroviruses can be identified.

1.3.2 Immunofluorescence

A specific antibody is conjugated (combined) with a fluorescent dye, fluoresceinisothiocyanate. The labeled antibody is then combined with the viral antigen. The complex formed is fluorescent and may be observed with a fluorescence microscope. The reaction is seen as a yellowish-green light.

There have been attempts to use the fluorescent antibody (FA) technique to detect and also quantify poliovirus in environmental samples, namely, tapwater. With this method it is possible to detect the presence of poliovirus in tissue culture within 6 to 9 hours instead of the 3 to 4 days normally required for visible observation of CPE in host cells. The quantitative determination of poliovirus necessitates, however, 18 to 24 hours of incubation, but it gives results comparable to those obtained by the conventional plaque-counting technique. The FA technique is summarized in Table 1.4.

TABLE 1.4.   Fluorescent-Antibody Technique for the
Rapid Detection and Measurement of Poliovirus 1
Numbers in Environmental Samples[a]

$4\text{–}5 \times 10^6$ host cells + poliovirus (+ tragacanth gum).

↓

Dispense mixture on microscope slides and incubate
at 37°C for 18–24 hours.

↓

Rinse slides and dry by acetone washing.

↓

Stain slides with rabbit antipolioglobulins labeled with
fluoresceinisothiocyanate.

↓

Examine slides under a fluorescence microscope.

---

[a] Adapted from E. Katzenelson (1976), *Arch. Virol.* **50**:197–206.

1.3.3   Enzyme Immunoassay

The enzyme immunoassay diagnostic procedure was first described in 1971 and
is known as the enzyme-linked-immunosorbent assay (ELISA). It can be used
to detect both viruses (antigens) and antibodies. Viruses are detected by the
double antibody sandwich method illustrated in Figure 1.9 and summarized as
follows: A solid surface is coated with a specific antibody and incubated at 4°C.
After washing, the antigen (virus) is added, and the mixture is incubated for
some hours at room temperature. After washing again, an enzyme-labeled
immunoglobulin containing a specific antibody is added to the mixture. The
solid surface is then incubated for 3 hours at room temperature and washed
again. Then enzyme substrate is added. This substrate may be 4-nitrophenyl
phosphate if the enzyme is alkaline phosphatase or 5-aminosalicyclic acid and
hydrogen peroxide for the enzyme peroxidase. When the reaction is stopped the
rate of substrate degradation is determined by spectrophotometry. This rate is
proportional to the amount of virus present in the sample.

ELISA may be used for the screening of viral diseases and thus may be suitable for epidemiological investigations. It has been successful in the detection of
hepatitis B antigen, adenoviruses, herpes simplex, coxsackieviruses, and many
other viruses. There is some hope that in the near future this simple and relatively safe assay may become routine in viral diagnostic laboratories.

1.3.4 Radioimmunoassay

Radioimmunoassay (RIA) is one of the most sensitive techniques presently available for the diagnosis of viral antigens or antibodies. This assay is based on the binding of an antigen by a specific antibody. The antigen is quantified by labeling the antibody with a radioisotope (^{125}I) and measuring the radioactivity bound to the antigen–antibody complex. The bound labeled virus is then separated from the mixture by centrifugation or filtration (in *liquid-phase RIA*) or by rinsing (in *solid-phase RIA*). The major steps involved in solid-phase RIA are illustrated in Figure 1.10. This technique has been used successfully for the detection of hepatitis A and hepatitis B antigens. Current techniques require 10^4 to 10^5 virus particles for detection by RIA.

It is unfortunate that ELISA and RIA do not distinguish between live and non-live virus particles.

1.3.5 Affinity Chromatography

Affinity chromatography consists of passing water through columns packed with beads that have been covered with specific antibodies. Viruses are trapped by the antibodies and are released afterward by leaching the column with sodium iodide. The essential features of this procedure are summarized in Table 1.5.

TABLE 1.5. Affinity Chromatography for the Isolation of Hepatitis Viruses from Water: Methodology[a]

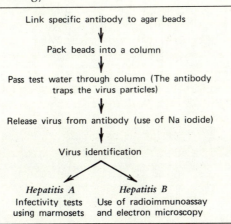

Link specific antibody to agar beads

Pack beads into a column

Pass test water through column (The antibody traps the virus particles)

Release virus from antibody (use of Na iodide)

Virus identification

Hepatitis A	*Hepatitis B*
Infectivity tests using marmosets	Use of radioimmunoassay and electron microscopy

[a] Adapted from W. O. K. Grabow and O. W. Prozesky (1977), In: *Int. Conf. Adv. Treat. and Reclam. Wastewater,* Johannesburg, South Africa.

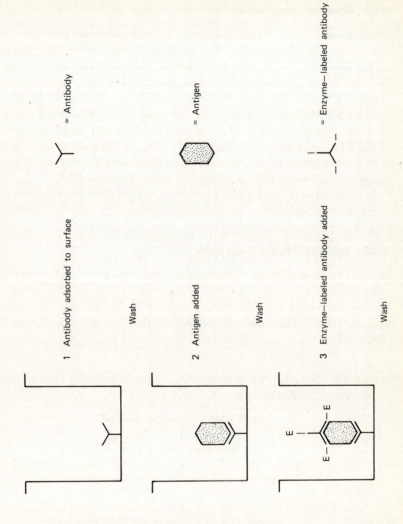

= Antibody

= Antigen

= Enzyme—labeled antibody

1 Antibody adsorbed to surface

Wash

2 Antigen added

Wash

3 Enzyme—labeled antibody added

Wash

E

E

E

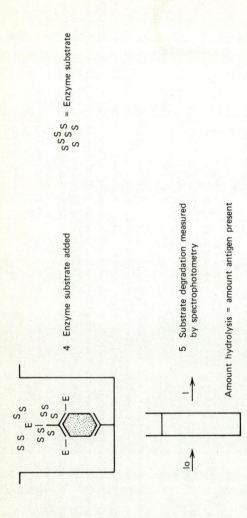

FIG. 1.9. Enzyme immunoassay: Double antibody sandwich method. Adapted from A. Voller et al. (1976), *Bull. WHO* **53**:55.

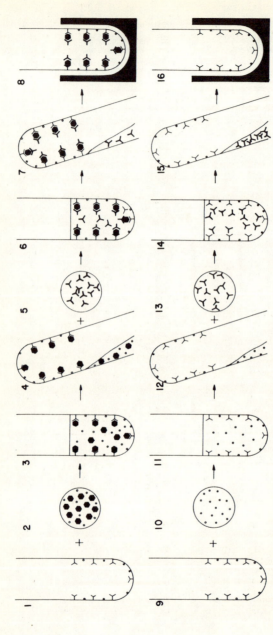

FIG. 1.10. Solid radioimmunoassay. Antibody (γ-shaped forms) specific for the desired antigen is absorbed onto a solid support, in this case the wall of a plastic test tube (1), along with nonspecific proteins (open circles). The material to be tested (2) is added; if it contains the antigen (black hexagons), the antigen combines with the antibody (3). Unattached antigen is washed out (4). To detect the bound antigen more antibody, now labeled with radioactive isotope (thickened γ-shaped forms), is added (5). It binds in turn to the antigen, which is in effect sandwiched between antibody layers (6). The tube is washed again, leaving only the antibody–antigen–antibody sandwiches (7). The radioactivity of the tube is measured in a radiation counter (8) and is compared with that derived from testing a negative control sample (9–16) to quantify the antigen. From J. L. Melnick et al. (1977), Sci. Am. **237**:44.

20

Affinity chromatography has been used for the selective isolation of viruses, namely, hepatitis B and possibly hepatitis A viruses, from 1 liter of water. There is hope that this method can be applied to other viruses and to larger volumes of environmental waters.

1.3.6 Immunoelectron Microscopy

Immunoelectron microscopy (IEM) is a technique by which fecal specimens or any other source of antigen are incubated with specific antibodies. After centrifugation the resulting pellet is resuspended and examined by electron microscopy for the presence of virus particles aggregated by the antibody.

This technique has proved to be successful in the examination of hepatitis viruses and the Norwalk agent.

1.4 CLASSIFICATION OF ANIMAL VIRUSES

DNA and RNA viruses can be distinguished by the viral nucleic acid core. Tables 1.6 and 1.7 give the major virus groups classified according to the capsid symmetry, presence of an envelope around the capsid, site of capsid assembly, reaction to ether, and diameter of the virion. There are five and eight major groups of DNA and RNA viruses, respectively. All DNA viruses have double-stranded DNA except members of the parvovirus group, and all RNA viruses have single-stranded RNA except members of the reovirus group.

1.4.1 DNA Viruses

Parvoviruses. Parvoviruses are very small (18–26 nm). The first parvovirus found in humans was the *adeno-associated virus* (AAV), which is responsible for childhood respiratory illness. It has been proposed that parvoviruses also may be responsible for acute infectious nonbacterial gastroenteritis.

Papovaviruses. In humans papovaviruses are responsible for *warts,* which are benign tumors. Papovavirus cause tumors in rabbits (rabbit papilloma virus) and other animals (polyoma virus or simian virus 40).

Adenoviruses. Adenoviruses are considered as part of the enteric virus group because they are shed in large numbers in feces. They cause respiratory illness and conjunctivitis (eye infection). They can be isolated from the adenoids, tonsils, conjunctiva, and respiratory or gastrointestinal tracts. Some of the agents cause tumors (sarcomas) in experimental animals, but there is no evidence that they do so in human hosts.

TABLE 1.6. Major Groups of Animal DNA[a] Viruses

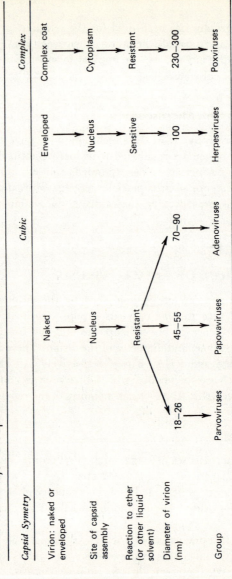

Capsid Symetry	Cubic			Complex
Virion: naked or enveloped	Naked	Enveloped		Complex coat
Site of capsid assembly	Nucleus	Nucleus		Cytoplasm
Reaction to ether (or other liquid solvent)	Resistant	Sensitive		Resistant
Diameter of virion (nm)	18–26 / 45–55 / 70–90	100		230–300
Group	Parvoviruses / Papovaviruses / Adenoviruses	Herpesviruses		Poxviruses

[a] All DNA viruses of vertebrates have double-stranded DNA, except members of the parvoviruses, which have single-stranded DNA. Adapted from J. L. Melnick (1976), *Prog. Med. Virol.* **22:**211–221.

TABLE 1.7. Major Groups of Animal RNA[a] Viruses

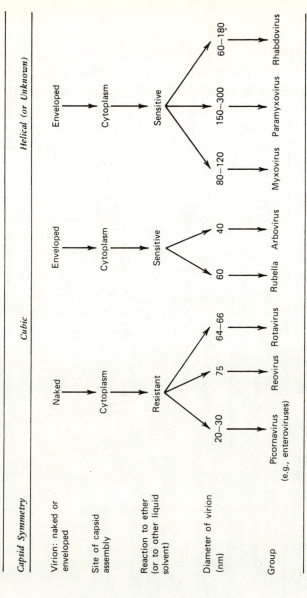

Capsid Symmetry	Cubic					Helical (or Unknown)		
Virion: naked or enveloped	Naked			Enveloped		Enveloped		
Site of capsid assembly	Cytoplasm			Cytoplasm		Cytoplasm		
Reaction to ether (or to other liquid solvent)	Resistant			Sensitive		Sensitive		
Diameter of virion (nm)	20–30	75	64–66	60	40	80–120	150–300	60–180
Group	Picornavirus (e.g., enteroviruses)	Reovirus	Rotavirus	Rubella	Arbovirus	Myxovirus	Paramyxovirus	Rhabdovirus

[a] All RNA viruses of vertebrates have single-stranded RNA, except members of the reovirus group, which are double-stranded. Adapted from J. L. Melnick (1976), *Prog. Med. Virol.* **22**:211–221.

TABLE 1.8. Some Characteristics of Human Enteric Viruses

Virus Group	Number of Types	Size (nm)	Type of Nucleic Acid	Ether Resistance	Some Diseases Caused	Site of Multiplication in the Human Body
Enterovirus 1. Poliovirus	3	20–30	RNA	Yes	Paralysis Aseptic meningitis	Spinal cord Lymphoid tissue Intestinal mucosa Meninges
2. Coxsackievirus A	24	20–30	RNA	Yes	Herpangia Aseptic meningitis Respiratory illness Paralysis Fever	Mouth Meninges Respiratory tract
B	6	20–30	RNA	Yes	Pleurodynia Aseptic meningitis Pericarditis Myocarditis Congenital heart anomalies Nephritis	Intercostal muscles Meninges Respiratory tract Developing heart

				Diseases	Tissue/Organ
3. Echovirus	34	RNA	Yes	Respiratory infection Aseptic meningitis Diarrhea Pericarditis Myocarditis Fever, rash	Respiratory tract Meninges Myocardial tissue Skin
Reovirus	3	RNA	Yes	Respiratory disease Gastroenteritis	
Adenovirus	31	DNA	Yes	Acute conjunctivitis Acute upper respiratory illness Eye infection	Conjunctival cells Respiratory tract
Hepatitis virus	3?	RNA?	?	Infectious hepatitis	Liver
	27 (Hep. A virus)				
	42 (Hep. B virus = Dane particle)	DNA	?	Serum hepatitis	Liver
Rotavirus	1?	RNA	Yes	Infantile gastroenteritis	Intestinal mucosa
Norwalk agent	1	DNA?	?	Nonbacterial gastroenteritis	Intestinal tract

Poxviruses. Poxviruses are the largest known animal viruses (approximately 200 to 250 nm). They cause diseases in humans (smallpox) and in animals (cowpox, fowlpox, monkeypox).

Herpesviruses. Herpesviruses cause latent infections in animals and in humans and do not tend to be important water contaminants.

1.4.2   RNA Viruses

Myxoviruses and Paramyxoviruses. Myxoviruses and paramyxoviruses are predominantly respiratory viruses that infect humans and animals. These two groups include influenza, parainfluenza, mumps, measles, and Newcastle disease viruses. Members of the myxovirus group have been isolated from lake water and sediment and may survive long enough to be transmitted by water.

Rhabdoviruses. The hosts of rhabdoviruses include mammals, insects, fishes, and plants. The *rabies virus* belongs to this group.

Arboviruses. Arboviruses are arthropod-borne (mosquitoes, ticks). Of 300 known arboviruses only 40 are capable of producing disease, such as:

- *Encephalitis.* "Equine encephalitis" virus attacks humans and equines.
- *Yellow Fever.* This disease causes muscle pain and rashes.
- *Dengue Fever.* It is the most widespread of the arboviral diseases, especially in Asia, the Pacific, and the Caribbean.

Enteroviruses and Reoviruses. Enteroviruses and reoviruses, referred to as *enteric viruses,* are examined in the following section and in Chapter 2.

1.5   ENTERIC VIRUSES

Enteric viruses are able to enter in the body via the oral route. They multiply in the cells of the gut and are excreted in large numbers in feces. For example, rotaviruses can be found in concentrations of 10^{10} particles per gram of feces of infected individuals. There are more than 100 types of enteric viruses. Some characteristics of human enteric viruses are given in Table 1.8.

Enteroviruses. Enteroviruses include polioviruses (3 serotypes), coxsackieviruses (24 serotypes of coxsackievirus A and 6 serotypes of coxsackievirus B), and echovirus (34 serotypes).

Reoviruses. There are three serotypes of reoviruses.

Adenoviruses. There are 31 serotypes of adenoviruses.

Hepatitis Viruses. Acute infection of the liver is caused by at least three viral agents (Type A, B and non-A, non-B viral hepatitis).

Rotavirus. The International Committee on Taxonomy of viruses has placed rotavirus in the reoviridae family. It is the main etiologic agent of infantile gastroenteritis throughout the world.

Norwalk Agent. The Norwalk agent is responsible for cases of gastroenteritis in adults and children. There is a current controversy about whether the Norwalk agent virus is a parvovirus (DNA core) or a picornaviruses (RNA core).

Enteric viruses are responsible for many diseases ranging from paralysis, meningitis, respiratory illnesses, congenital heart anomalies, and conjunctivitis, to diarrhea, fever, and skin rash. Most cases of infections due to enteric viruses do not result in overt diseases that are clinically detectable. On the other hand, this may explain why their waterborne dissemination has been poorly documented. This subject is discussed in Chapter 2.

1.6 SUMMARY

Viruses are colloidal infectious particles ranging in size from 20 to 350 nm. They are composed of a protein coat (capsid) containing a nucleic acid core. Some viruses have an envelope surrounding the protein coat.

Viruses are classified according to a number of characteristics, including the type of nucleic acid, host cell, size, capsid architecture, presence or absence of envelopes, and susceptibility to chemical agents.

Viruses are obligate intracellular parasites that reproduce only inside an appropriate host cell. Their lytic cycle includes six steps: adsorption to and penetration into host cells, eclipse, replication, maturation, and release of mature virions.

Tissue cultures are the most popular for the cultivation and enumeration of enteric viruses.

A number of procedures are used for viral diagnosis. These procedures include neutralization tests, immunofluorescence, enzyme immunoassays, and radioimmunoassay. There is an urgent need to develop rapid diagnostic tests for environmental monitoring of viruses.

1.7 FURTHER READING

Bradley, D. E. 1967. Ultrastructure of bacteriophages and bacteriocins. *Bacteriol. Rev.* **31**:230–314.

Dalton, A. J., and F. Haguenau. 1973. *Ultrastructure of Animal Viruses and Bacteriophages.* Academic, New York.

Fox, J. P. 1974. Human-associated viruses in water. In: *Viruses in Water,* G. Berg et al., Eds. American Public Health Association, Washington, D.C.

Hsiung, G. D. 1973. *Diagnostic Virology: An Illustrated Handbook.* Yale University Press, New Haven, Conn.

Kalter, S. S., and C. H. Millstein. 1974. Animal associated viruses. In: *Viruses in Water,* G. Berg et al., Eds. American Public Health Association, Washington, D.C.

Kurstak, E., and C. Kurstak, Eds. 1977. *Comparative Diagnosis of Viral Diseases,* Vol. 2. Academic, New York.

Lemke, P. A., and C. H. Nash, 1974. Fungal viruses. *Bacteriol. Rev.* **38**:29–56.

Lennette, E. H., and N. J. Schmidt, Eds. 1969. *Diagnostic Procedures for Viral and Rickettsial Infections,* 4th ed. American Public Health Association, Washington, D.C.

Melnick, J. L. 1976. Taxonomy of viruses. *Prog. Med. Virol.* **22**:211–221.

Padan, E., and M. Shilo. 1973. Cyanophages. Viruses attacking blue-green algae. *Bacteriol. Rev.* **37**:343–370.

Prier, J. E. 1966. *Basic Medical Virology.* Williams and Wilkins, Baltimore, Md.

Rovozzo, G. C., and C. N. Burke. 1973. *Basic Virological Techniques.* Prentice-Hall, Englewood Cliffs, N.J.

Sigel, M. M., D. G. Rippe, A. R. Beasley, and M. Dorsey, Jr. 1976. Systems for detecting viruses and viral activity. In: *Viruses in Water,* G. Berg et al., Eds. American Public Health Association, Washington, D.C.

two
Epidemiology

The object of this chapter is to familiarize the reader with some essential epidemiological concepts and to discuss the epidemiology of major enteric viral diseases.

2.1 ELEMENTS OF EPIDEMIOLOGY

2.1.1 Some Definitions

Epidemiology is the study of the factors that influence the frequency and distribution of infectious diseases in man. It is aimed at identifying the etiology of the disease, host factors associated with the disease, and the influence of environmental factors on the disease process.

The continuing presence of a disease within a geographical area is called *endemic*, but when the occurrence of the disease is much above the normal expectancy it is called an *epidemic*. *Pandemic* refers to the spread of the disease across continents. With regard to diseases of animals the words *epizootic* and *enzootic* are equivalent to epidemic and endemic, respectively.

2.1.2   Chain of Infection

There are six main factors governing the infectious disease process:

- Infectious agent.
- Reservoir.
- Portal of escape.
- Transmission.
- Portal of entry.
- Susceptibility of the host.

Infectious Agent

There is a wide variety of infectious agents that cause diseases in man. These agents include bacteria, fungi, protozoa, metazoa, rickettsiae, and viruses. Once the infectious agent has entered the host, several results may follow (Figure 2.1):

Colonization. Colonization is the establishment of the parasite within the host.

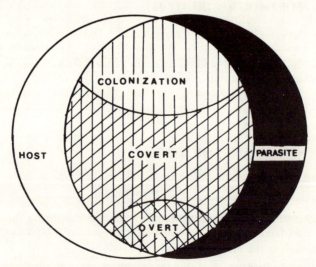

FIG. 2.1.   Host–parasite interactions. From J. S. Mausner and A. K. Bahn (1974), *Epidemiology: An Introductory Text*, W. B. Saunders Co., Philadelphia, Pa. Courtesy of W. B. Saunders Co.

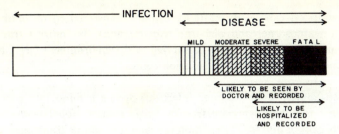

FIG. 2.2. Relation of severity of illness to disease statistics. From J. S. Mausner and A. K. Bahn (1974), *Epidemiology: An Introductory Text,* W. B. Saunders Co., Philadelphia, Pa. Courtesy of W. B. Saunders Co.

Inapparent Infection or Covert Infection. Infection is the entry and development of the infectious agent inside the body. The host reacts to the agent, but the reaction is not clinically detectable.

Overt Disease. Disease is the development of clinical symptoms that are easily detectable. The development of the disease depends on various factors, including infectious dose, virulence (i.e., pathogenicity or ability to induce disease), and environment (temperature, humidity, radiation, prior host exposure to agent). Infectious diseases may vary in severity. In some cases, the infectious process may result in severe or fatal illness (e.g., rabies virus), whereas in others it may lead mostly to inapparent infection (e.g., poliovirus and other enteric viruses). Since most of the enteric viruses cause inapparent infection, it is generally difficult to keep any accurate records of the diseases they cause. Figure 2.2 shows the relation of the severity of the illness to disease statistics.

Reservoir
A reservoir is the natural habitat where the infectious agent normally lives and multiply. The reservoir for viruses is generally a living organism such as humans or arthropods. A person can acquire the infectious agent from another person or from animals (zoonoses).

Portal of Escape
Viruses generally escape from the infected host through the respiratory or alimentary tracts. Many viruses are known to be heavily excreted in feces, and this is a general characteristic of the enteric virus group.

Transmission
Transmission or spread involves the transport of the infectious agent from the reservoir to the host. It is the most important link in the chain of infection.

The transmission can be *direct* (e.g., contact transmission) or *indirect* (e.g., foodborne, waterborne, and airborne transmission). In indirect transmission the infectious agent must be able to survive in the environment (see Chapters 3 and 4). There are four main transmission routes.

Common-Source Transmission. Infectious agents coming into contact with susceptible hosts may originate from a single source. Foodborne and waterborne outbreaks of disease are examples of common-source transmission.

Direct or Contact Transmission. Disease can be transmitted through contact between the source and the susceptible host or via droplet generation by coughing or sneezing within a few feet of the host.

Airborne Transmission. The airborne route of transmission involves dust particles and microbial aerosols. Aerosols can remain airborne for long distances before reaching a susceptible host. Aerosol particles smaller than 5 μm are of greatest concern because they are able to reach the alveoli and subsequently induce disease.

Vectorborne Transmission. Insects are involved in the transmission of some viral diseases. It has been suggested that they may also play a role in the transmission of enteric viruses.

Portal of Entry

The mouth is the major portal of entry of many enteric viruses. Other viruses may enter through the upper respiratory tract.

Susceptibility of the Host

The susceptibility of the host to infectious agents depends on a variety of factors, which include the following:

1. *Host or Intrinsic Factors.* Specific immunity, age, sex, and so on.
2. *Environmental or Extrinsic Factors.* Diet, temperature, humidity, radiation, socioeconomic level, hygienic conditions, and so on.

We shall now examine the epidemiology of the main viral diseases and discuss their transmission by the water route.

2.2 EPIDEMIOLOGY OF VIRAL DISEASES

Person-to-person contacts constitute the main mode of transmission of enteric viral infections and diseases. Small children are generally the most susceptible group within a community. Their close contact with other children and their unhygienic habits predispose them to higher rates of infection. Adults are generally less subject to infection because of potential acquired immunity during previous contacts with the virus. The infection depends on the hygienic and socioeconomic levels of the population and on the season (enteric virus outbreaks are more prevalent during summer and early fall).

We have briefly reviewed the general pattern of transmission of enteric viruses. In addition to the person-to-person mode of transmission any enteric virus may theoretically be transmissible by the water route. However, this waterborne transmission has been definitely demonstrated only for the virus causing infectious hepatitis. The difficulties may be because of our inability to detect the numerous inapparent infections caused by enteric viruses and the fact that waterborne transmission is sometimes masked by person-to-person contact spread. Viruses are shed in large quantities in feces of healthy *carriers* (a carrier is an infected individual who does not show any symptoms or signs of infection), and their number in raw municipal sewage may range from 10^3 to 10^6 PFU/l. These viral pathogens may survive waste and water treatment operations and come into contact with man via food, water, or aerosols generated by spray irrigation with wastewater effluents. The potential modes of transmission of enteric viruses are illustrated in Figure 2.3.

We will now discuss the epidemiology of some important viral diseases.

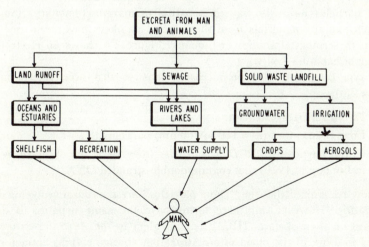

FIG. 2.3.   Possible modes of enteric virus transmission. From C. P. Gerba et al. (1975), *Environ. Sci. Technol.* **9**:1122–1126. Courtesy of American Chemical Society.

2.2.1   Viral Hepatitis

Viral hepatitis has been recognized as a major health problem for centuries. It is traditionally associated with epidemics in army installations. Numerous outbreaks have been reported in U.S. military bases in Germany during World War II.

The most characteristic symptom of viral hepatitis is jaundice, which is a yellow discoloration of the skin and the white of the eyes. This disease is caused by three viruses:

- *Type A Viral Hepatitis.* It is also named infectious hepatitis or short-incubation hepatitis (15–40 days).

- *Type B Viral Hepatitis.* It is also called serum hepatitis or long-incubation hepatitis (50–180 days).

- *Non-A, Non-B Viral Hepatitis.* There are some reasons to believe that a third viral agent causes hepatitis, which is also associated with blood handling.

The major features of human hepatitis A and B are shown in Table 2.1.

The development of sophisticated diagnostic procedures (e.g., IEM) has allowed the visualization of the viruses responsible for hepatitis A and B. The agent for hepatitis A was first observed in 1973 in stools of volunteers who were administered serum from hepatitis A patients. The enterovirus-like particles are 27 nm in diameter and contain a genome probably made of RNA. These particles have also caused hepatitis in marmoset monkeys (*Saguinus mystax*), and the materials from the infected monkeys have been used as an antigen in serological studies in humans. Figure 2.4 shows an electron micrograph of hepatitis A virus.

For type B hepatitis electron microscopy has revealed three types of particles (Figure 2.5).

- 22-nm spherical particles.
- Long filamentous ($L = 100$ to 700 nm) particles.
- 42-nm particles called *Dane particles* (they were first observed by Dane in London in 1970) that contain double-stranded DNA.

It is now recognized that the "Dane particles" are the causative agents of type B hepatitis. However, all three forms have the same hepatitis B antigen (HB_sAg) on their surface. HB_sAg is equivalent to the *Australia antigen* discovered in 1965 in the blood of an Australian aborigine. This antigen can be detected easily by RIA. The "Dane particle" contains also a *core antigen* or HB_cAg. This antigen induces the formation of an antibody (anti-HB_cAg) that is released much earlier than the one induced by HB_sAg and may serve for early diagnosis of type B hepatitis.

TABLE 2.1. Main Features of Human Hepatitis A and B[a]

Feature	Hepatitis A	Hepatitis B
Name	Infectious hepatitis = short incubation hepatitis	Serum hepatitis = long incubation hepatitis
Incubation period	15–40 days	50–180 days
Mortality	Low	High
Antigenicity	No cross-immunity to hepatitis B virus	No cross-immunity to hepatitis A virus
Antigens in blood	Not usually present	Present during incubation period and acute phase
Virus	Enterovirus-like 27 nm particle (RNA virus?)	42-nm double-shelled DNA virion (Dane particle)
Shedding of virus in feces	Demonstrated	Not demonstrated
Epidemiology	Fecal–oral route via direct contact or food and waterborne transmission	Via transfer of blood products and via direct contact

[a] Adapted from R. F. Boyd and B. G. Hoerl (1977), *Basic Medical Microbiology*, Little, Brown and Co., Boston, Mass.; and from A. P. Waterson (1976), *Ann. Rev. Med.* **27**:23–35.

Since hepatitis viruses cannot be cultured and identified by available tissue culture techniques, subclinical cases of hepatitis can be documented only by serological tests (e.g., RIA) and also by biochemical tests aimed at monitoring changes in enzyme systems such as the increase in alanine aminotransferase and aspartate transaminase.

In 1976 there were 60,000 reported cases of acute hepatitis in the United States: 60% of these cases were attributed to hepatitis A, 25% to hepatitis B, and 15% were unspecified. Type A hepatitis classically occurs in the age group of 1 to 15 years and becomes less prevalent beyond the age of 35. It has a low fatality rate of 0.5% as compared to 3–4% with type B hepatitis.

Type A hepatitis is transmitted by the fecal–oral route by several mechanisms:

Person-to-Person Contact. The major mode of transmission of type A hepatitis is person-to-person contact.

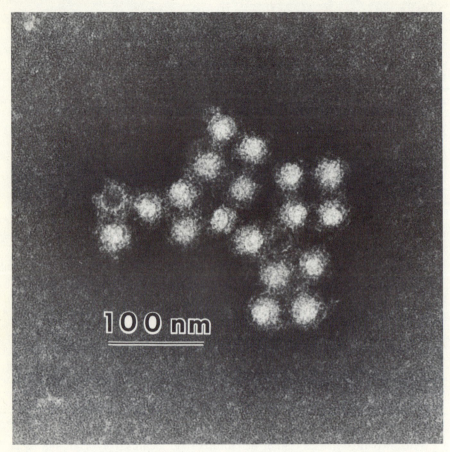

FIG. 2.4.   Electron micrograph of hepatitis A virus (chimpanzee fecal extract). Courtesy of C. R. Gravelle.

Waterborne Transmission. Hepatitis A virus was recognized as truly waterborne and can be transmitted via consumption of improperly treated water. There have been many reported outbreaks associated with contaminated drinking water supplies (Table 2.2), and the most important one is undoubtedly the one that occurred in Delhi, India, that resulted in approximately 30,000 cases. This unfortunate episode showed that chlorination was probably unable to inactivate the virus responsible for this disease. It is, however, important to note that swimming in contaminated water has not been definitely associated with hepatitis A infection.

Foodborne Transmission. Hepatitis A virus can be transmitted by consumption of shellfish grown in sewage-contaminated waters, contamination of food by infected handlers, and use of contaminated water for washing containers (see Chapter 11).

Hepatitis B is transmitted through the use of contaminated needles and syringes (e.g., drug addicts), blood and blood products, hemodialysis units, dental procedures, and tatooing. It is also an occupational disease that may strike physicians, oral surgeons, dentists and workers in blood transfusion centers. Hepatitis B virus can also be spread by intimate physical contact (e.g., sexual transmission). Its waterborne transmission is unlikely, since hepatitis B antigen is not excreted in feces.

Hepatitis A can be prevented by adhering to strict sanitary conditions. Gamma globulin may be administered to hepatitis patients to prevent or at least attenuate the disease. For type B hepatitis blood donors must be screened for hepatitis B antigen and equipment must be properly disinfected. There is now some hope that a vaccine against hepatitis B virus can be routinely used to prevent the disease. This vaccine is made of surface antigens found on the 22-nm particles that occur abundantly in the blood of apparently healthy carriers.

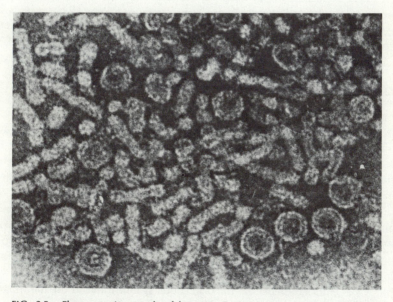

FIG. 2.5. Electron micrograph of hepatitis B virus. Note the three different particles: Filamentous particles, small (22 nm) particles, and double-walled "Dane particles." Courtesy of G. A. Cabral and J. L. Melnick.

TABLE 2.2.   Some Outbreaks of Hepatitis Associated with Drinking Water (1947–1976)[a]

Date of Occurrence	Place	No. of Cases	Type of Water Supply
1947–1948	Germany (U.S. Army)	52	Municipal system
1948–1949	Sweden	391	Municipal system
1952	USA (summer camp)	102	Deep well
1955–1956	Delhi, India	28,745	Municipal system
1966	France	427	Public supply
1969	Massachusetts (Holy Cross College)	90	Public supply
1971	Tennessee	129	River water
1971	Alabama	50	Semipublic supply
1976	China	>1000	Well water

[a] Adapted from J. W. Moseley (1965), in: *Transmission of Viruses by the Water Route*, G. Berg, Ed., Wiley, New York; M. Goldfield (1974), in: *Viruses in Water*, G. Berg et al., Eds. APHA; and from *WHO Scient. Group on Human Viruses in Water, Wastewater and Soil*, Geneva, 23–27 October 1978.

2.2.2   Poliomyelitis

Poliomyelitis is a disease known to the civilized world for many centuries. In 1894 a polio epidemic in Vermont resulted in 132 cases with a death rate of 13.5%. One of the most ravaging epidemics occurred in New York City in 1913 with 9000 cases. The mystery of this dreadful disease stirred the scientific world in Europe and in the United States, and in 1908 Karl Landsteiner reported that poliomyelitis was caused by a "filterable agent." Examination of this infectious agent was possible only after the discovery of the electron microscope in 1931. The major breakthrough in poliomyelitis research occurred in 1949. Enders and his group succeeded in cultivating poliovirus *in vitro* on tissue cultures. Since that significant achievement, the way to a vaccine against poliovirus was open.

Poliovirus enters the host via the oral route. It multiplies in the nasopharynx and on the intestinal mucosa, and then passes into the bloodstream. It is shed abundantly in the feces during the incubation period, which lasts from 10 to 15 days. Fortunately, epidemiological evidence indicates that only 1 in 100 persons infected by the virus shows clinical symptoms of the disease. Paralytic poliomyelitis results from the invasion of the central nervous system by the viral agent. Paralysis affects mainly adolescents and young adults.

In 1953 there was another breakthrough in polio research. Jonas Salk proposed a vaccine which was made of inactivated poliovirus types 1, 2, and 3. In 1955 Albert Sabin proposed a vaccine composed of live but attenuated strains of the three serotypes of poliovirus. This vaccine can be taken orally, usually on sugar cubes. The U.S. Public Health Service licensed the Sabin vaccine in 1962. Since the introduction of the vaccines the paralytic poliomyelitis cases have dramatically decreased from 13,850 cases/year in 1955 to less than 35 cases/year in 1969–1972. There have been reports, however, of polio cases associated with the administration of vaccines, but they are usually rare.

Public health reports from Sweden in the 1940s attributed a great importance to water in the transmission of poliovirus. However, the only documented case of waterborne transmission was the poliovirus epidemic in Huskerville, Nebraska, in 1952. Of 45 cases 17 showed paralytic poliomyelitis. Therefore, although waterborne transmission of poliovirus is a distinct possibility, person-to-person contact remains the prevailing way for the spread of the disease.

2.2.3 Gastroenteritis

Gastroenteritis is a disease characterized by abdominal pain, diarrhea, headache, and fever. The disease lasts from 1 to 3 days. In children, acute gastroenteritis may be fatal, and death may occur within 48 hours. The major factors causing death are electrolyte imbalance and dehydration, leading to cardiac arrest. From May 1972 to March 1977, there were 21 fatal cases of *infantile acute gastroenteritis* in Toronto, Canada. This disease largely contributes to childhood mortality in developing countries, and it is estimated that this disease is responsible for 5 to 18 million childhood deaths per year in Africa, Asia, and Latin America.

It is estimated that bacteria are involved in 20 to 30% of gastroenteritis cases, whereas viruses account for approximately 50% of the cases. With respect to viruses, some strains of echoviruses, coxsackieviruses, adenoviruses, and the Norwalk agent (27-nm particles, which probably are parvoviruses) have been linked to gastroenteritis. More recently, electron-microscopy investigations have revealed that a rotavirus belonging to the reovirus group (see Table 1.8) is the major cause of gastroenteritis in children. Figure 2.6 shows an electron micrograph of a simian rotavirus (SA11) that grows well in tissue culture. Rotavirus was indeed involved in the 21 fatal gastroenteritis cases that occurred in Toronto, Canada. The rotavirus gastroenteritis is prevalent during the colder months of the year.

Public water, semipublic water, ice, shellfish, shipboard water, and lake water have been implicated in the transmission of gastroenteritis (Table 2.3). In the most recent (1978) gastroenteritis outbreak in Vermont 2000 cases— 20% of a town's population—were affected by the disease. The outbreak was associated with the consumption of contaminated water.

2.2.4   Diseases Caused by Other Enteric Viruses

Other enteric viruses, coxsackie- and echo- (=enteric cytopathogenic human orphan) viruses are responsible for mostly subclinical infections but may also be the cause of acute diseases. They have an RNA core and range in size from 20 to 30 nm (Table 1.8). There are 30 strains of coxsackieviruses (24 strains of coxsackie A and 6 strains of coxsackie B) and 34 strains of echoviruses. They multiply in the alimentary tract and pass into the bloodstream before entering the central nervous system to cause *meningitis*. They are also the cause of *herpangina* (infection of the throat), respiratory symptoms, *myalgia* (muscular pain), *exanthem* (eruption or rash on the skin), *nephritis* (inflammation of the kidneys), *pleurodynia* (pain of the intercostal muscles or of the pleural nerves), congenital heart anomalies, and also *myocarditis* (inflammation of the myocardium) and *pericarditis* (inflammation of the pericardium, the sac enveloping the

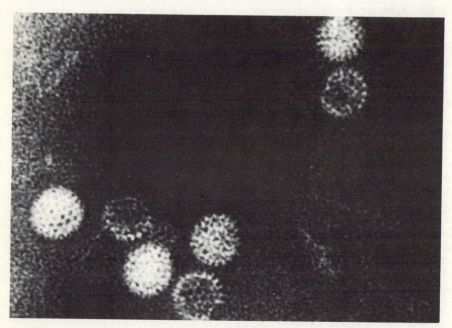

FIG. 2.6.   Simian rotavirus (SAII) purified from MA-104 monkey kidney cells. Magnification 213,120. Courtesy of E. M. Smith.

TABLE 2.3.   Some Waterborne Outbreaks of Gastroenteritis[a]

Date of Occurrence	Place	Vehicle	Number of Cases
August 1965	California	Public water	2,500
June 1966	New Jersey and Pennsylvania	Semipublic water	13,000
January 1967	New York	Shellfish	7
September 1967	West Virginia	Ice	700
May 1972	California	Lake water	26
April 1973	Wisconsin	Shipboard water	26
June 1978	Vermont	Public water	2,000

[a] Adapted from M. Goldfield (1974), in: *Viruses in Water*, G. Berg, H. L. Bodily, E. H. Lennette, J. L. Melnick, and T. G. Metcalf, Eds., American Public Health Association.

heart). A World Health Organization report revealed that, during a 5-year period (1969–1973), 300 of 1295 cases of cardiac diseases were caused by coxsackie B viruses.

These viruses are distributed worldwide and readily infect children. The infection rates are greatest during the summer months under temperate climates. They are spread by direct contact and possibly by respiratory droplets and insects. They have sometimes been associated with waterborne epidemics, but there is no definite proof of their transmission by this route.

Adenoviruses are the only enteric viruses that contain a DNA core. There are 31 strains of adenoviruses that range in size from 68 to 85 nm. Adenoviruses are respiratory viruses that also can infect the alimentary tract and thus be excreted in large quantities by infected individuals. Swimming pools are known to act as a means of transmission of these viruses, resulting in conjunctivitis outbreaks.

2.3   THE QUESTION OF MINIMAL INFECTIVE DOSE

For bacterial pathogens one single cell is rarely enough to cause infection. The minimal infective dose is generally high. For instance, 10^5 cells of *Salmonella* must be consumed to produce disease in healthy male volunteers (Table 2.4). Surprisingly, the infective dose for *Shigella* is relatively low (10–100 cells) as compared to *Salmonella* sp. or *Vibrio cholerae*. Therefore, 10 to 100 *Shigella* cells remaining viable on the clothing or hands of infected patients may be easily transmitted and cause bacillary dysentery.

A little more than a decade ago it was generally believed that large number of viruses were necessary to trigger the infection process. Today one accepts the laboratory findings that one infective dose of tissue culture (1 PFU) is sufficient to infect humans. One must remember that infection and disease are not synonymous. An infected person does not show any overt recognizable signs of the disease. However, the infection process can be detected by demonstration of viruses in the host or by observation of the interaction of the virus with a specific antibody, using sophisticated techniques such as IEM. It has been estimated that 1 in 100 to 1000 infected persons will manifest the symptoms of the disease. Unfortunately, infected persons may act as inapparent carriers and subsequently help spread the disease within the community through contact with healthy individuals. It is concluded that the presence of even small numbers of viruses in drinking water supplies should be prevented.

2.4   ANIMAL-ASSOCIATED VIRUSES

Many viruses of animal origin (cattle, swine, cats, dogs, fish, etc.) may be excreted in large numbers and find their way into our waterways, including water supplies. The implications of their presence in water have not been evaluated. In cattle feed lots the concentration of animals in small areas greatly facilitates the interchange of these viruses within a group of animals. Of greater importance is the question of interchange of viruses between humans and animals. There are some data that support the association of human viruses with animal hosts.

It has been suggested that there is also the possibility of interchange of viruses between host belonging to aquatic and terrestrial ecosystems. Racoons, among others, generally interact with both terrestrial (humans, domestic and wild animals) and aquatic (crustacea) hosts for viruses. The hypothetical viral

TABLE 2.4.   Infective Dose of Some Bacterial Enteric Pathogens[a]

Bacteria	Infective Dose
Shigella	10^1–10^2
Salmonella	10^5
Escherichia coli	10^8
Vibrio cholerae	$<10^2$

[a] H. L. DuPont and R. B. Hornick (1973) Med. **52**:262–270; E. R. Eubanks et al. (1976) Infect. Immun. **13**:457–463

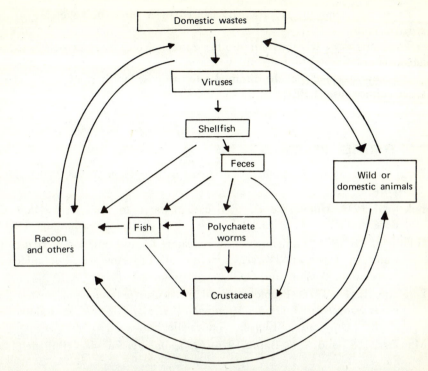

FIG. 2.7. Hypothetical viral transmission routes between terrestrial and estuarive ecosystems. From T. G. Metcalf (1976), in: *Virus Aspects of Applying Municipal Wastes to Land,* L. B. Baldwin *et al.,* Eds. University of Florida, Gainesville, Fla.

transmission between terrestrial and aquatic ecosystems is illustrated in Figure 2.7.

2.5 SUMMARY

The major mode of transmission of enteric viral disease is by person-to-person contact.

The waterborne and foodborne transmission of type A hepatitis has been demonstrated. The agent causing this disease is a 27-nm enterovirus-like particle which contains a genome probably made of RNA. The virus cannot be cultivated by tissue culture techniques and subclinical cases can be demonstrated only by serological and biochemical tests.

Waterborne transmission of poliovirus is a distinct possibility, and contact transmission is the major mode of the spread of poliomyelitis.

Acute gastroenteritis is primarily caused by a rotavirus belonging to the reovirus group.

The waterborne transmission of other enteric viruses (coxsackie-, echo-, and adenoviruses) has not been definitely proven.

It is now accepted that the minimal infective dose for viruses is one infective tissue culture dose (1 PFU).

2.6   FURTHER READING

Berg, G., Ed. 1967. *Transmission of Virus by the Water Route*, Wiley, New York.

Fox, J. P. 1974. Human associated viruses in water. In: *Viruses in Water*. G. Berg et al., Eds. American Public Health Association, Washington, D.C.

Goldfield, M. 1974. Epidemiological indicators for transmission of viruses in water. In: *Viruses in Water*, G. Berg et al., Eds. American Public Health Association, Washington, D.C.

Lennette, E. H. 1976. Problems posed by viruses in municipal wastes. In: *Virus Aspects of Applying Municipal Waste to Land*. L. B. Baldwin et al., Eds. University of Florida, Gainesville, Fla.

Mausner, J. S., and A. K. Bahn. 1974. *Epidemiology*, Saunders, Philadelphia, Pa.

Melnick, J. L., G. R. Dreesman, and F. B. Hollinger. 1977. Viral hepatitis. *Sci. Am.* **237**:44–52.

Moseley, J. W. 1965. Transmission of viral diseases by drinking water. In: *Transmission of Viruses by the Water Route*, G. Berg, Ed. Wiley, New York.

Steinhoff, M. C. 1978. Viruses and diarrhea: A review. *Am. J. Dis. Child.* **132**:302–307.

Top, F. H. Sr., and P. F. Wehrle, Eds. 1976. *Communicable and Infectious Diseases*, Mosby, St. Louis, Mo.

three
Survival of Viruses in the Environment:
Influence of Physical, Chemical, and Biological Factors on Virus Persistence

3.1   INTRODUCTION

Viruses are obligate intracellular parasites that can multiply only inside a susceptible host cell. Once in the host cell-free state, they are exposed to a variety of adverse environmental factors of physicochemical (temperature, light, desiccation, pH) and biological (enzymes, polysaccharides, etc.) nature. Their stability under these environmental stresses is an important parameter in their successful transmission from one host to another.

In Chapter 1, we examined the structure of viruses which are essentially composed of the following:

- A protein coat.
- A nucleic acid core.
- A lipid envelope for some viruses.

Environmental factors may exert a specific action on one of these viral components. *Inactivation* is a process by which viruses lose their ability to produce progeny. Following treatment with some chemicals (i.e., formaldehyde) viruses can lose their *infectivity* (irreversible destruction of their nucleic acid) without destruction of their antigenic properties. In this case the protein coat is not denatured and is able to trigger an immune response from the host individual. Other chemicals, such as phenol, destroy the capsid without altering the nucleic acid core. The extracted nucleic acid can infect host cells, although with reduced efficiency.

The loss of infectivity should normally follow the first-order kinetics according to the equation

$$\frac{N}{N_0} = e^{-kt} \tag{3.1}$$

where t = time
N = virus infectivity at time t
N_0 = Initial virus infectivity
k = rate constant

A straight line should be obtained when the logarithm of the virus titer is plotted against time. However, according to the degree of virus clumping or to other factors, many types of survival curves are obtained. Figure 3.1 illustrates the shape of the survival curves obtained with virus particles. In practical terms, any inactivation test should be undertaken with:

- Well-dispersed virus particles.
- A virus suspension free of extraneous substances which might protect the virus from the environmental agent under investigation.

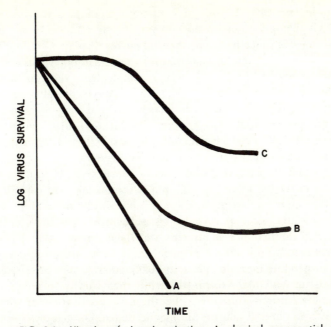

FIG. 3.1. Kinetics of virus inactivation. A, classical exponential survival curve. Most of the viruses occur as single particles. B, A high percentage of viruses occur as single particles. The few remaining are in clumps. C, Viruses occur as aggregated particles. Adapted from S. L. Chang (1967), in: *Transmission of Viruses by the Water Route*, G. Berg, Ed., Wiley-Interscience, New York.

We will now examine separately the effect of physical, chemical, and biological factors on virus inactivation in the cell-free state.

3.2 PHYSICAL INACTIVATION OF VIRUSES

Physical factors probably play the most significant role in the fate of viruses in the cell-free state and in their transmission from one host to another. The most important physical agents are temperature, light, and desiccation.

3.2.1 Temperature

Temperature is probably the most active physical factor with respect to disruption of viruses outside their host cells. Thermal inactivation is widely used by humans to eliminate viruses, primarily from food products. Most viruses are inactivated in less than 1 hour at temperatures ranging from 55 to 65°C. However, virus populations do not always display uniformity as far as their thermoresistance is concerned. The thermal curves display a "tailing effect," as shown in Figure 3.1. Some thermoresistant variants of poliovirus can survive a temperature as high as 75°C. However, recent evidence suggests that the thermoresistant viruses are probably infectious RNA released after capsid breakdown. Exposure to high temperatures may result in damage to both the nucleic acid and the protein coat. The protein denaturation affects virus infectivity through prevention of virus adsorption to the host cells and through prevention of replication due to the inactivation of specific enzymes.

It appears that at relatively "high" temperatures (e.g., 60°C), the viral protein coat is inactivated more rapidly than is the nucleic acid core. It has been suggested that an oxidation process was involved in the thermal inactivation of some enteroviruses, and that reducing substances (cysteine) decrease the damage due to thermal stress. Proteins and some cations (Mg^{2+}) may protect viruses from thermal disruption. Heat inactivation is further influenced by the pH of the suspending medium. Thus, at 50°C, enteroviruses are protected from heat by NaCl or $MgCl_2$ at neutral pH, while their survival is reduced at pH = 3.1. Moreover, anions can affect the thermal inactivation of viruses. For example, at 50°C and at neutral pH magnesium sulfate has a lower protective effect than magnesium chloride.

The above data concern the thermal inactivation of viruses suspended in well-defined media. It is now accepted that temperature is the most significant factor controlling virus survival in the more complex environments such as freshwater, seawater, soil, or sludge. However, the mechanism of enhancement or reduction in virus survival in these environments is not very clear. For example, it was found that clay minerals or soil can provide some degree of protection from heat. At temperatures ranging from 43 to 51°C raw sludge also has a protective effect toward poliovirus as compared to survival in digested sludge or phosphate buffered saline (see Chapter 10).

Viruses are not inactivated at very low temperatures. In fact they are commonly stored at −70°C in protein solutions (i.e., fetal calf serum).

3.2.2   Light

Light, especially its ultraviolet (UV) portion has long been known for its germicidal properties. Ultraviolet nonionizing radiation, in conjunction with chemicals, has long been used in the sterilization of vaccines. The site of inactivation by UV light is the viral nucleic acid. The lethal radiation affects the pyrimidine rings in the nucleic acid with the subsequent formation of thymine dimers (i.e., formation of covalent bonds between adjacent thymine residues). Figure 3.2A shows the formation of a thymine dimer following UV irradiation.

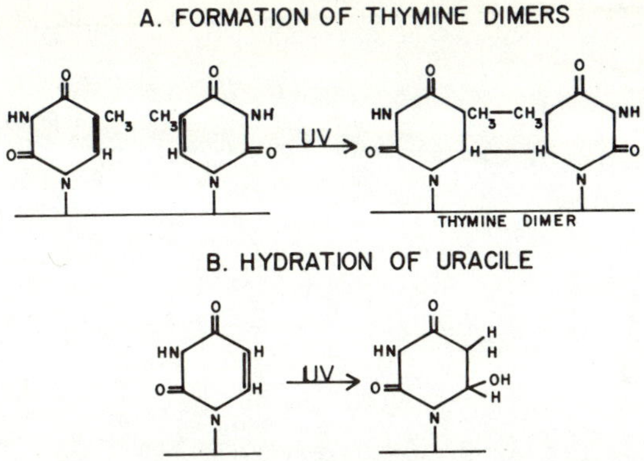

A. FORMATION OF THYMINE DIMERS

B. HYDRATION OF URACILE

FIG. 3.2. Damage to nucleic acid following UV irradiation. From C. A. Knight, 1975. *Chemistry of Viruses*. Springer-Verlag, New York.

Damage can also be due to the hydration of uracil residues. This is illustrated in Figure 3.2B. As compared to ionizing radiation (x rays, gamma rays), UV action toward microbes is very selective. The lethal action of UV is usually reduced by turbidity and color present in the suspending medium. For many years UV radiation has been proposed for the disinfection of water supplies and sewage effluents and for shellfish depuration in lieu of chlorination. This process would have the advantage of producing effluents with no toxic residuals, as in the case of chlorination (i.e., chlorinated organics). However, this system has four major disadvantages:

- It does not leave any residual for protection from any potential post-contamination.
- The water must have a low turbidity and color.
- Problems arise when UV irradiation is applied to large scale operations.
- The process is more expensive than conventional disinfection.

This question will be further discussed in Chapter 7.

Much less information is available on the effect of visible light on virus inactivation. It has been shown that artificial visible light can inactivate some animal viruses (Newcastle disease virus, fowl plague virus, measles virus). Sunlight probably plays a rather important role in the inactivation of viruses in natural waters and in engineered systems such as oxidation ponds. The effect of sunlight on the destruction of an enterovirus is illustrated in Figure 3.3. In a water with low turbidity (in Jackson turbidity units: 1.7 JTU) 80% of poliovirus type 1 was inactivated within 3 hours when the mean light intensity was 0.646 cal cm^{-2} min^{-1} and the mean temperature was 26°C. In natural

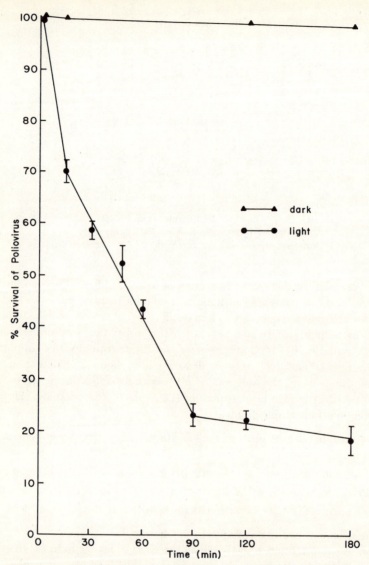

FIG. 3.3.   Effect of sunlight on poliovirus survival. From G. Bitton and R. Fraxedas (unpublished data).

waters, virus inactivation is probably more complex. The water may contain substances which reduce the photoinactivation of viruses. This topic is discussed in Chapter 4, which deals with the fate of viruses in the aquatic environment.

Most enteric viruses are very sensitive to visible light in the presence of oxygen and dyes. This process is called *photodynamic inactivation*. The dye is a heterocyclic compound such as methylene blue or neutral red. The virus inactivation results from a photooxidation process catalyzed by the dye that attaches to the nucleic acid. The light source can be a flow-through irradiation cell illustrated in Figure 3.4 or solar radiation. Many factors control this inactivation process:

- pH.
- Dye concentration.
- Sensitization time.
- Light intensity.
- Temperature.

Figure 3.5 shows the decline of poliovirus under optimal conditions. A significant virus destruction was observed within a 5-minute period.

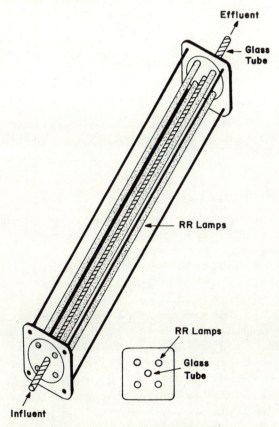

FIG. 3.4.   Flow-through irradiation cell. From C. P. Gerba et al. (1977), *J. Water Poll. Control Fed.* **49:**575–583. Courtesy of Water Pollution Control Federation.

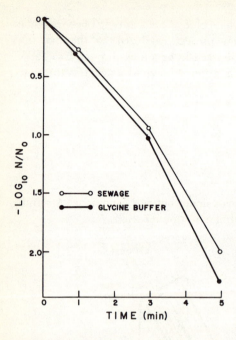

FIG. 3.5. Photodynamic inactivation of poliovirus in wastewater and in glycine buffer. (Both samples were adjusted to pH 10.0 and contained 5 mg/l methylene blue and were held for 24 hours in the dark before light exposure. The turbidity was 5 JTU after adjustment to pH 10.0.) From C. P. Gerba et al. (1977), *J. Water Poll. Control Fed.* **49**:575–583. Courtesy of Water Pollution Control Federation.

Although this process is quite efficient in destroying viruses, there are still some unanswered questions about the health effects of residual dye that may remain in the effluents.

3.2.3   Desiccation

Desiccation is an important factor controlling the persistence of viruses on inanimate objects (laboratory surfaces, clothing, dust particles), in aerosols, in soils, and in sludges applied to drying beds.

The persistence of viruses on inanimate surfaces is of importance in the transmission of viruses within the hospital environment. Some viruses can persist for long periods on common surfaces such as glass slides, ceramic tiles, vinyl floors, and stainless steel surfaces. Table 3.1 illustrates the survival of enteric viruses on glass slides at low (7%) and high (96%) relative humidity (RH). Adenovirus 2 and poliovirus 2 survive for at least 8 weeks at 7% RH, whereas coxsackievirus B3 could survive for only 2 weeks on that particular surface. Enteroviruses may also display a long survival (up to 20 weeks) at 35% RH on the surface of clothing material. The survival is generally higher on wool than on cotton fabrics.

Relative humidity is probably the most decisive factor controlling the survival of viruses in the aerosol state. This subject is reviewed in more detail in Chapter 9.

It has been demonstrated that desiccation plays an important role in virus inactivation in sludges and in soils. In soils rapid inactivation occurs when the moisture content is below 10%, whereas in sludge a virus decline is observed when the concentration of solids is above 65%. In both cases viral inactivation is due to the release and the subsequent breakdown of the RNA core.

A practical application of these phenomena is the problem of preservation of enteric viruses by freeze-drying. The use of a suitable suspending medium (i.e., Tris buffer) devoid of salts can result in the survival of most enterovirus groups during the drying process. This is of practical importance to laboratories concerned with the preservation and shipping of virus samples.

3.2.4   Inactivation of Viruses at the Air–Water Interface

Some 30 years ago straightforward experiments had shown that viruses can be inactivated at the air–water interface. This interface is easily increased by shaking or by bubbling air through a virus suspension. The surface inactivation of many viruses was investigated and it was found that some viruses like T1, MS2, and SFV (Semliki forest virus) are rapidly inactivated, whereas others, such as encephalomyocarditis virus, an enterovirus, remain unaffected. The surface inactivation of phage MS2 by shaking is illustrated in Figure 3.6A. The virus decline increases with the salt concentration of the suspending medium. High salt concentration brings about a better adsorption of the virion to the air–water interface and thus its rapid decline. The presence of viruses at this

TABLE 3.1.   Influence of Relative Humidity on the Survival of Enteric Viruses on Glass Slides at 25°C[a]

Virus	Relative Humidity (%)	Number of Viruses/Slide[b]			
		4 hours	1 day	2 weeks	8 weeks
Adenovirus 2	96	$10^{7.5}$	$10^{8.25}$	$10^{6.0}$	$10^{2.75}$
	7	$10^{7.5}$	$10^{7.5}$	$10^{7.0}$	$10^{5.25}$
Poliovirus 2	96	8.0×10^7	1.2×10^8	6.0×10^6	NVD[c]
	7	1.6×10^6	1.1×10^6	3.2×10^5	2.0×10^2
Coxsackievirus	96	$10^{6.25}$	$10^{5.25}$	$10^{2.5}$	NVD[d]
	7	$10^{4.75}$	$10^{4.5}$	$10^{2.5}$	NVD

[a] Adapted from M. C. Mahl and C. Sadler (1975), *Can. J. Microbiol.* **21**:819–823.

[b] Numbers in $TCID_{50}$/ml (adenovirus and coxsackievirus) or in PFU/ml (poliovirus).

[c] No virus detected.

[d] No virus was detected after 3 weeks.

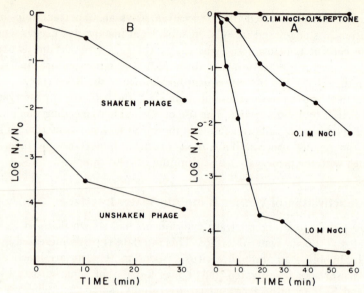

FIG. 3.6. (A) Inactivation of phage MS2 by shaking. (The stock phage suspension was diluted 1:10⁴ in the indicated media and then subjected to surface inactivation by shaking 10 ml of suspension in a 30-ml bottle by means of a flask shaker.) (B) Inactivation in aerosol at 90% RH of the fraction of phage MS2 surviving in a shaking experiment. (A sample of phage was subjected to shaking in 1 M NaCl until a more resistant fraction appeared and then sprayed. As a control a nonshaken sample was sprayed from 1 M NaCl.) From T. Trouwborst and J. C. DeJong (1973), *Appl. Microbiol.* **26:**252–257. Courtesy of American Society for Microbiology.

interface is also influenced by the nature of their surface components. Viruses with a lipid envelope tend to adsorb more easily to the air–water interface where they are more rapidly inactivated. The affinity of these viruses for the air–water interface is due to the hydrophobic nature of their envelope. On the other hand, proteins and surface active agents protect viruses from inactivation at the air–water interface (Figure 3.6A). These substances act by preventing the entry of viruses into the air–water interface.

Surface inactivation plays a significant role in the viability of viruses in the aerosol state. Thus viruses resistant to shaking (i.e., surface inactivation) tend to survive better in the airborne state (Figure 3.6B)

3.2.5 Pressure

Practically nothing is known about the effect of pressure on viral function. This information would be most useful in assessing the fate of pathogenic viruses following deep-sea disposal of human wastes (depths ranging from 2500 to 5000 m). We know that indicator bacteria such as *E. coli* and *Streptococcus faecalis* survive the rigors of pressures ranging from 250 to 500 atm at 4°C, and one could assume that viruses would survive to the same extent as bacteria.

3.2.6   Sonic Energy

Viruses are generally stable when exposed to ultrasonic treatment. However, long sonication periods can lead to some inactivation. An increase of the sonication time from 3 to 10 minutes can substantially reduce viral numbers (Table 3.2). Sonication is commonly used for breaking virus aggregates, releasing viruses embedded within sludge flocs or desorbing them from various surfaces (Table 3.2).

3.2.7   Aggregation of Viruses

It has long been suspected that virus aggregation influence their survival under environmental conditions. Extensive research at the University of North Carolina has shed some light on this important phenomenon. Sophisticated

TABLE 3.2.   Elution of Poliovirus Type 1[a] (Sabin) from Magnetite by Casein[b]

Treatment	Number of Viruses Eluted (PFU/ml)	% Elution
1 hour contact time with eluent	4,950	36.8
1 hour contact time with eluent + 3 minutes sonication[d]	12,350	89.2
1 hour contact time with eluent + 10 minutes sonication	2,650	19.1

[a] Adapted from G. Bitton and O. Pancorbo (unpublished). Poliovirus Type 1 was shaken from 20 minutes in the presence of 500 ppm magnetite and 1610 ppm $CaCl_2$ and the mixture was pelleted down to determine the virus adsorbed to magnetite. Under these conditions, 99.1% of the virus was sorbed by magnetite.

[b] 100 ppm casein, Tris buffered to pH = 9.

[c] 13,850 PFU/ml represent 100% elution.

[d] The mixtures were sonicated in an ice bath using the standard probe of a Branson Sonifier Model S-75 with a power output of 75 W and an ultrasonic frequency of 20 kHz.

techniques have been devised to assess the extent of viral aggregation in natural waters and to demonstrate the impact of this phenomenon on virus persistence. Viruses are released from infected host cells in a highly aggregated state and they probably remain so upon entering wastewaters and natural waters. Ionic conditions prevailing in natural waters do not lead to viral dispersion. Under experimental conditions viruses are aggegated at acidic (pH = 3 or 5), but not at alkaline pHs. Figure 3.7 shows aggregates of poliovirus (Mahoney strain) in

FIG. 3.7.   Aggregation of poliovirus (Mahoney strain) after dilution in 0.05 M glycine buffer at pH = 3.0. Courtesy of R. Floyd.

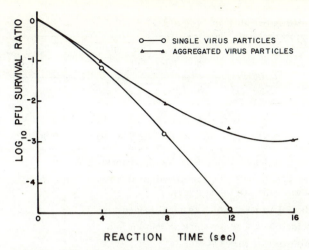

FIG. 3.8.   Effect of aggregation on the disinfection efficiency of bromine (10 μM). Adapted from D. C. Young and D. G. Sharp (1977), *Appl. Environ. Microbiol.* **33**:168–177.

0.05 M glycine buffer, at pH 3.0. Cations such as Mg^{2+} promote viral dispersion, whereas Ca^{2+} and Al^{3+}, under appropriate concentrations and in combination, were found to induce aggregation.

Furthermore, aggregation exerts an influence on virus survival in natural waters. Viruses within the aggregates are probably highly resistant to the destructive action of environmental factors. Moreover, aggregation influences the outcome of chemical disinfection of viruses. For example, Figure 3.8 reveals a strong influence of aggregation on chemical disinfection with bromine. Aggregated poliovirus particles are 2 orders of magnitude more resistant to bromine than dispersed virions.

3.2.8   Persistence of Viruses in Space

We have reviewed the main physical factors that may influence the survival of viruses in the environment. During space flights viruses would be subjected to some of these physical stresses: vibration, ultraviolet radiation, x-rays, cosmic rays, temperature fluctuations, and weightlessness.

Survival experiments (Table 3.3) carried out aboard a rocket, which reached a height of 155 km, have shown a 3 log reduction in poliovirus survival under flight conditions. However, T1 phage displayed a higher survival than poliovirus. This shows that the rigors of space flight can reduce the infectivity of some viruses.

TABLE 3.3.   Virus Survival in Space[a,b]

Virus	Total PFU Seeded	Ground Controls Total PFU		Flight[c] Samples Total PFU	
		Shielded	Exposed	Shielded[d]	Exposed[e]
Poliovirus 3	1.5×10^7	5.3×10^3	3.1×10^3	3.3×10^1	—
T1 phage	9×10^6	8.3×10^6	2.9×10^6	2.6×10^6	1×10^2

[a] Adapted from J. Hotchin et al. (1965), *Nature* **206**:442–445.

[b] Surfaces were coated with dried preparations of viruses. The survivors were counted at 26 days after coating of the surfaces.

[c] Flight was aboard a rocket which reached a height of 155 km.

[d] Samples shielded with a 38-μm thick aluminum foil.

[e] Samples were exposed for 233 seconds.

3.3   CHEMICAL INACTIVATION OF VIRUSES

Inactivation of viruses by chemical means is a significant tool used by humans in their constant fight against these infectious agents. In contrast to physical agents, chemicals probably play a minor role in viral inactivation in the natural environment. Therefore the emphasis will be placed on chemical substances deliberately introduced in the virus milieu to bring about their inactivation.

Prior to introducing the chemicals that inactivate viruses *in vitro*, one should remember that host cells have their own specific ways of fighting invasion by viruses; this process is called *in vivo inactivation*. Antiviral chemicals released *in vivo* are as follows:

- *Antibodies.* These are specific proteins produced by the host organism to fight foreign substances called *antigens*. In this case viruses are the antigens. The production of antibodies is triggered by vaccines (live or attenuated strains of viruses) or by natural infection.

- *Interferon.* This is a glycoprotein produced by the host to fight virus infection. It acts indirectly by inducing the production of an antiviral protein. The use of interferon therapy has not yet reached a practical stage.

We will now examine the chemical inactivation of virus *in vitro*.

3.3.1   pH

Although some viruses are unstable in a low pH environment, those of enteric origin are not susceptible to acidic conditions in general. However, high pH is detrimental to viruses and pH above 11 is commonly very destructive. Lime treatment of water and wastewater results in high pH values that effectively destroy viruses. The effect of high pH on virus decline is illustrated in Figure 3.9. The effect is not selective and may lead to changes in both the protein coat and nucleic acid. It is essential to remember that environmental factors can interact with each other. For example, we have already seen that pH can influence the heat stability of viruses. Some enteric viruses, particularly adenoviruses, are very sensitive to high pH values. They may be completely inactivated within 10 minutes at pH above 10. However, viruses are generally stable at pHs commonly encountered in natural waters (pH 5 to 9).

3.3.2 Chemicals That Alter the Nucleic Acid

Chemicals that alter the nucleic acid core are as follows:

1. *Formaldehyde.* This aldehyde has been used for many years in the production of vaccines because it does not alter the antigenic properties of the virus.
2. *Nitrous Acid.* Its reaction with the nucleic acid results in a deamination of purine and pyrimidine bases.
3. *Ammonia.* It is also able to cause cleavage of RNA within virus particles.

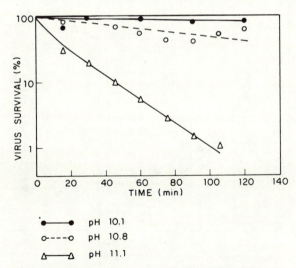

FIG. 3.9. Effect of high pH on poliovirus 1 (LSc). From G. Berg et al. (1968), *J. Am. Water Works Assoc.* **60**:193. Courtesy of American Water Works Association.

3.3.3   Chemicals That Alter the Protein Coat

Phenol is an example of a chemical substance that affects the proteinaceous capsid. It is commonly used in the isolation of infectious nucleic acid.

It has been recently shown that exposure of certain viruses to low ionic strength results in subtle changes in the protein coat, which prevent these viruses from adsorbing to their host cells. Within 30 to 40 minutes 99% reduction in infectivity is displayed by coxsackievirus A13 in a low ionic strength environment (Figure 3.10). This phenomenon is temperature-dependent and

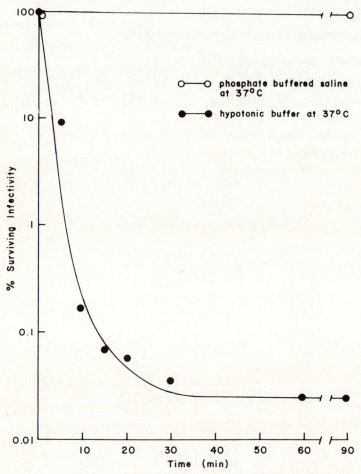

FIG. 3.10.   Inactivation of coxsackievirus A13 at low ionic strength. From Cords et al. (1975), *J. Virol.* **15:**244–252. Courtesy of American Society for Microbiology.

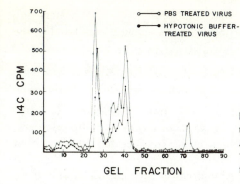

FIG. 3.11. Effect of low ionic strength on the polypeptide profile of coxsackievirus A13. From C. E. Cords et al. (1975), *J. Virol.* **15:**244–252. Courtesy of American Society for Microbiology.

concerns other types of coxsackie A viruses. However, poliovirus 1 and cox-sackie B viruses are not inactivated in low ionic strength conditions. The capsid of pircornaviruses is composed of four major structural polypeptides: VP1, VP2, VP3, and VP4 (decreasing molecular weight). The polypeptide involved in adsorption to host cells varies with the type of virus. The exposure of cox-sackievirus A13 to low ionic strength results in the loss of structural polypeptide VP4, and this is illustrated in the polypeptide profile of the virus under normal (phosphate buffered saline) and low ionic strength conditions (Figure 3.11).

3.3.4 Chemicals That Affect Lipid Envelopes

Ether and sodium dodecyl sulfate (SDS) are typical chemicals that alter the lipid envelope of some viruses such as influenza virus. In fact, one of the criteria used for classification of viruses is their resistance to ether. We have seen in Table 1.8 that all enteric viruses are ether-resistant. Ether is sometimes used for decontamination (selective destruction of bacterial and fungal contaminants) of viral suspensions.

3.3.5 Oxidizing Agents

Halogens (chlorine, bromine, iodine) and ozone are commonly used in the disinfection of water and wastewater. Their antiviral action is discussed in more detail in Chapter 8.

3.3.6 Heavy Metals

Heavy metals have been suspected as potential virucidal agents in seawater. The extended survival observed in autoclaved seawater may be due to the precipitation of heavy metals during that treatment. In natural waters any virucidal action of heavy metals may be influenced by changes in pH and by the complexing action of aquatic organic matter.

3.3.7   Antiviral Chemotherapy

We have seen that one can trigger antibody production with vaccines. Unfortunately, only a limited number of virus types (polio, measles, smallpox, rabies, yellow fever) can be controlled by vaccines. Hence considerable effort has been expended on finding antiviral substances that would alter some viral function without affecting the host cells. Among hundreds of such chemicals only three have been approved by the Federal Drug Administration (FDA):

- *Amantadine*. This is used against some strains of influenza and parainfluenza viruses.
- *Idoxuridine and Adenine Arabinoside* (Ara-A). These are used in the treatment of eye infections produced by herpes virus.

One realizes that antiviral chemotherapy is an important research field that should be directed toward a search for efficient and nontoxic chemicals.

3.4   BIOLOGICAL INACTIVATION OF VIRUSES

Major components of the biota have been studied for their interactions with viruses. Information about these interactions has expanded our knowledge of the fate of viruses in water and wastewaters and subsequently of the public health implications.

Presently there is no substantial evidence of a host–parasite relationship between mammalian viruses and bacteria, algae, protozoa, or certain metazoa. This means that animal viruses can multiply only inside a homologous host cell line. Therefore interaction of viruses with the biota can result only in no-effect or inactivation.

3.4.1   Interaction with Bacteria

Bacteria have been implicated in the decline of viruses in water and wastewater. Three types of bacteria have been shown to inactivate viruses in activated sludge systems. The virus decline may be due to enzymatic action. Table 3.4 shows the inactivation of coxsackievirus A9 by three bacterial species: *Bacillus subtilis, Escherichia coli,* and *Pseudomonas aeruginosa. B. subtilis* and *P. aeruginosa* displayed an antiviral activity, and the virus capsid was used as a substrate for bacterial growth.

TABLE 3.4. Net Inactivation ($-\log_{10}$) of Coxsackievirus A9 by 3 Bacterial Species[a]

Incubation Temperature	Time (days)	E. coli	B. subtilis	P. aeruginosa
24°C	4	0.01	0.25	1.53
	8	0.04	0.86	2.40
37°C	4	0	1.13	2.38
	8	0	1.29	2.00

[a] Adapted from D. O. Cliver and J. E. Herrmann (1972), *Water Res.* **6:**797–805.

An important feature of the transmission of enteric viruses is their general resistance to proteolytic enzymes commonly found in the intestinal tract. Coxsackievirus A9 is an exception, since its proteinaceous capsid is digested by trypsin and pronase. The effect of the latter is illustrated in Table 3.5. Proteolytic enzymes are used in the purification of enteroviruses from protein contamination. They may sometimes increase virus infectivity. For example, pepsin may reactivate poliovirus that has been neutralized by antibody, and chymotrypsin may enhance the infectivity of reoviruses. The mechanism of this enzymatic enhancement of infectivity is not quite clear.

TABLE 3.5. Effect of Pronase on Coxsackievirus A9[a]

Treatment	Infectivity (PFU/ml)
Virus control[b]	5.4×10^7
Pronase	2.0×10^4
PMSF-Pronase[c]	2.8×10^7

[a] Adapted from J. E. Herrmann and D. O. Cliver (1973), *Inf. Immun.* **7:**513–517.

[b] Virus suspended in phosphate buffered saline.

[c] Pronase inhibited by phenylmethyl sulfonylfluoride (PMSF). The virus control for this experiment was 3.4×10^7 PFU/ml.

3.4.2 Algae

It has been known for many years that microscopic and macroscopic algae have certain antibiotic properties. The chemicals released by algae vary from acrylic acid to polyphenols. This, of course, has a great impact on the biological equilibrium within the aquatic environment.

Algal activity (photosynthesis) may indirectly influence bacterial and possibly viral stability by increasing the dissolved oxygen and the pH of the water. During an algal bloom the pH of the water can be as high as 10, and we have seen that viruses can be more easily inactivated on the alkaline than on the acidic side. This phenomenon has some impact upon virus survival in oxidation ponds.

It has been known for almost 20 years that polysaccharides extracted from Rhodophyta (red algae) can protect chicken embryos from influenza B and mumps virus infection. More recently, other members of the Rhodophyta have displayed similar antiviral activity toward herpes simplex 1 and 2 and toward an enterovirus, coxsackievirus B5. It has been hypothesized that these algal polysaccharides inhibit the adsorption of viruses to the host cell. Dextran sulfate, a synthetic sulfated polysaccharide, can also interfere with the multiplication of herpes, encephalomyocarditis (EMC), and poliovirus. For some viruses (i.e., poliovirus) the inhibitory effect of dextran depends on its molecular weight. The effect of this polysaccharide on these three types of viruses is shown in Table 3.6.

TABLE 3.6.   Effect of Dextran Sulfate on Herpes, EMC, and Polio Viruses[a]

Dextran Sulfate[b] MW ($\times 10^4$)	Plaque Inhibition of		
	Herpes	EMC	Poliovirus
2–3	+[c]	−	−
6–8	+	+	−
25–30	+	+	−
50	+	+	+
200	+	+	+

[a] Adapted from K. K. Takemoto and O. Spicer (1965), *Ann. N.Y. Acad. Sci.* **130**:365–373.

[b] Dextran sulfate added at a concentration of 0.2 mg/ml of overlay medium.

[c] + = inhibition; − = no effect.

3.4.3 Protozoa

Some protozoa belonging to the Ciliata (e.g., *Tetrahymena pyriformis*) or Sarcodina (e.g., *Naegleria gruberi*) groups are able to ingest enteroviruses. The viruses are subsequently inactivated within the protozoa. However, this type of interaction is probably infrequent.

3.4.4 Animals

The oceans have been most examined for the presence of antiviral substances. In fact two of the most used antiviral substances, Ara-A and cytosine arabinoside (Ara-C), were isolated from marine sponges. As mentioned earlier, these chemicals are useful in the treatment of herpetic infections.

Marine animals, namely, abalones and clams, contain antiviral substances. The active agent extracted from clams is a glycoprotein called *paolin*. The conch mucus has also been implicated in the inhibition of infectivity of poliovirus type 3. However, the mechanism of this inhibition is not known yet.

3.5 INTERACTION OF VIRUSES WITH SURFACES

Viruses are considered as biocolloids with electrical properties governing their adsorption to biological and nonbiological surfaces. Their surface electrical charges result from the ionization of carboxyl (COO^-) and amino (NH_3^+) groups localized on the surface of their protein coat. Once in the cell-free state, viruses may interact with solid surfaces, and this association is due to Brownian motion, electrostatic forces, and electrical double-layer phenomena. Furthermore, the association is governed by the physical and chemical properties of the suspending medium, namely, the cation concentration, pH, and organic matter. The attachment of viruses to suspended solids in water and wastewater has many implications in the survival of viruses in the cell-free state. In appropriate chapters of this book more details will be given on the importance of this phenomenon on the survival of viruses within the soil, sediments and aquatic environments, and on the chemical disinfection of water and wastewater.

3.6 SUMMARY

Viruses are subject, outside their host cells, to inactivation by physical, chemical, and biological factors.

Among the physical factors, temperature is the most significant in causing virus decline. Viruses probably enter natural waters in an aggregated state and this may increase their survival under environmental stresses.

In comparison to physical factors, chemicals probably play a minor role in viral inactivation in the natural environment. However, chemical means are used by human beings in their constant fight against these infectious agents. The field of antiviral chemotherapy is expanding and efforts are being made to search for efficient and nontoxic chemicals.

Viruses are also subject to inactivation by numerous components of the biota, and there is no evidence to support viral multiplication outside a homologous host cell line.

3.7   FURTHER READING

Bitton, G. 1978. Survival of enteric viruses. In: *Water Pollution Microbiology*, Vol. 2. R. Mitchell, Ed. Wiley-Interscience, New York.

Bitton, G. 1979. Adsorption of viruses to surfaces: Technological and ecological implications. In: *Adsorption of Microorganisms to Surfaces*, G. Bitton and K. C. Marshall, Eds. Wiley-Interscience, New York.

Chang, S. L. 1970. Interactions between animal viruses and higher forms of microbes. *J. San. Eng. Div.* **96**:151.

Knight, C. A. 1975. *Chemistry of Viruses.* Springer-Verlag, New York.

Marshall, K. C. 1976. *Interfaces in Microbiol Ecology.* Harvard University Press, Cambridge, Mass.

Morris, E. J., and H. M. Darlow. 1971. Inactivation of viruses. In: *Inhibition and Destruction of the Microbial Cells.* W. B. Hugo, Ed. Academic, New York.

Walker, B., Jr. 1970. Viruses respond to environmental exposure. *J. Environ. Health* **32**:532–550.

four
Fate of Viruses in the Aquatic Environment

4.1 INTRODUCTION

We have discussed in detail the environmental factors controlling the survival of viruses in the cell-free state. In the last decade water has been recognized as a possible route for the transmission of enteric viral diseases. We have now some evidence that infectious hepatitis virus may be transmitted via the water route (see Chapter 2). However, there is a lack of epidemiological data concerning the association between swimming and viral diseases. Recently it has been demonstrated that there was a statistically significant association between swimming and shigellosis. *Shigella sonnei* was isolated from the Mississippi River near Dubuque, Iowa, and was shown to be responsible for disease in individuals who had water in their mouths while swimming in the river. The attack rates versus the degree of exposure to water is shown in Table 4.1. These are significant data, since *Shigella* has a low infectious dose (10 to 100

TABLE 4.1.   Shigellosis: Swimmers' Attack Rates
Versus the Degree of Exposure to Water[a]

Degree of Exposure to Water	% Ill
In river but not wet above waist	0
Wet above waist but head not wet	0
Head wet but not under water	0
Head under water but no water in mouth	3
Water in mouth	18

[a] Adapted from M. L. Rosenberg, K. K. Hazlet, J. Schaefer, J.
  G. Wells, and R. C. Pruneda (1976), *JAMA* **236**:1849.

bacterial cells) as compared with other pathogens (see Chapter 2). This
infectious dose is not very different from that of enteric viruses. A recent study
has shown an increased risk of contracting enteric diseases for those bathers who
actually immerse their heads into contaminated seawater. Thus the transmission of viral diseases via swimming is indeed possible, but unfortunately the
epidemiological picture is complicated by the fact that enteric viruses produce a
wide range of subclinical diseases. Some concern has also been raised over the
adsorption of viruses onto bubbles in the sea surf and their transport, in the
aerosolized state, to public beaches.

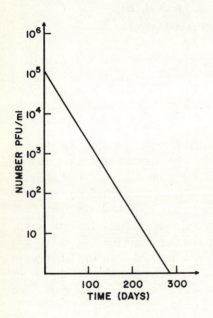

FIG. 4.1.   Survival of poliovirus 1 in a well-
defined medium at 18–24°C. (Adapted from L.
Schwartzbrod et al., (1975), *Rev. Epidemiol.
Med. Soc.* **23**:235–252.)

Defined aqueous medium:

NaHCO$_3$	275 mg/l
CaCl$_2$	101 mg/l
MgSO$_4$·H$_2$O	233 mg/l
HCl (N)	0.5 ml/l
Distilled water	1000 ml
pH = 6.9	

TABLE 4.2.   Municipal Waste Discharges in the Coastal Zone, Including Estuaries[a]

Region[b]	Total Volume of Municipal Wastes (mgd)[c]	Sewered Population with Secondary Treatment 1968 (%)[d]
North Atlantic	550	25
Middle Atlantic	3500	60
Chesapeake Bay	640	90
South Atlantic	370	75
Caribbean	160	—[e]
Gulf of Mexico	760	75
Pacific Southwest	1900	35
Pacific Northwest	390	50
Alaska	13	25
Pacific Islands	85	25
Total	8300	50

[a] From U.S. Department of the Interior (1970), National Estuarine Pollution Study, U.S. Government Printing Office. *Reference:* J. K. McNulty (1977), in: *Coastal Ecosystem Management,* J. R. Clark, Ed., John Wiley and Sons, New York.

[b] North Atlantic: Canadian border to Cape Cod.
Middle Atlantic: Cape Cod to Cape Hatteras excluding Chesapeake Bay.
Chesapeake Bay: Chesapeake Bay systems inland of Cape Charles and Cape Henry.
South Atlantic: Cape Hatteras to Fort Lauderdale, FL.
Caribbean: Fort Lauderdale to Cape Romano, FL., plus Puerto Rico and Virgin Islands.
Gulf of Mexico: Cape Romano to Mexican border.
Pacific Southwest: Mexican border to Cape Mendocino.
Pacific Northwest: Cape Mendocino to Canadian border.
Alaska: Alaska.
Pacific Islands: Hawaii, Guam, American Samoa.

[c] Based on 150 gallons per capita per day of total population in coastal counties plus a few noncoastal counties, 1965.

[d] Data from U.S. Department of the Interior, Federal Water Pollution Control Administration, 1969.

[e] Not available.

Since viruses may cause disease in people swimming in or consuming shellfish from contaminated areas, we are obviously interested in their survival in the aquatic environment. Potentially, enteric viruses can survive for long periods in well-buffered aqueous media of known chemical composition. Thus poliovirus type 1 can survive up to 300 days in these media (Figure 4.1). However, aquatic environments are not as well-defined chemically, and viruses entering water will be exposed to a wide range of physical, chemical, and biological factors. The purpose of this chapter is to examine the behavior of viruses in fresh and marine waters, and in sediments.

4.2   FATE OF VIRUSES IN SEAWATER

Wastes resulting from man's activities have been traditionally (and are still) discharged into the ocean. These include liquid wastewater, sludge, oil, and industrial and thermal wastes. With ever-increasing populations along the coastlines around the world it is now realized that the ocean can no longer be considered as the "*infinite sink*" for the disposal of fecal and other wastes. The coastal waters in the United States receive more than 8 billion gallons of municipal sewage in a single day (Table 4.2). Approximately half of this domestic sewage receives secondary treatment.

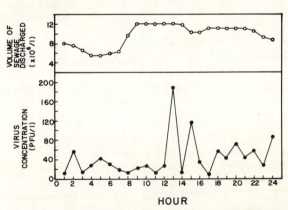

FIG. 4.2.   Discharge of Honolulu's raw sewage into Mamala Bay. (O—O) volume of sewage discharged over a 24-hour period; (●—●) concentrations of enteroviruses recovered from sewage collected by hourly grab sample method over a 24-hour period. From G. G. Ruiter and R. S. Fujioka (1978), *Water, Air Soil Poll.* **10**:95–103. Courtesy of D. Reidel Publ. Co., The Netherlands.

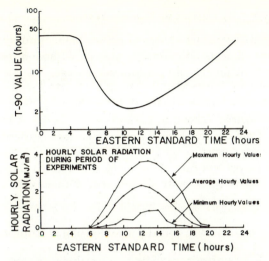

FIG. 4.3. Effect of solar radiation on coliform bacteria in seawater. From: J. T. Bellair et al. (1977), *J. Water Poll. Control Fed.* **49**:2022–2030. Courtesy of Water Pollution Control Federation.

Ocean outfalls are commonly used for the disposal of nontreated or partially treated sewage into the ocean. Their main advantage is low cost as compared to tertiary treatment of wastewater. Their design is based on location, water depth (for example, the west coast of the United States has deeper waters than the Atlantic coast), distance from shore, and mixing regime. The outfall pipe may extend from 1 to 4 miles offshore and is from 2 to 12 feet in diameter.

Researchers at the University of Hawaii have shown that up to 8×10^{10} viruses are discharged daily into Mamala Bay via a sewage ocean outfall. Hourly fluctuations in virus concentration (15–195 PFU/l) were observed during a 24-hour period with a peak concentration occurring between 11:00 A.M. and 2:00 P.M. (Figure 4.2).

Over the past 25 years much information has been gathered concerning the fate of enteric bacteria in ocean water. These pathogens are generally eliminated by physical dilution and by die-off mechanisms, which include temperature, predation, nutrient deficiency, algal and bacterial antibiotics or toxins, solar radiation, heavy metals, and salinity. Other factors, such as sedimentation, lead to their accumulation in marine sediments. It recently has been found that solar radiation plays a significant role in the die-off of coliforms in seawater. Figure 4.3 shows that the T_{90} (time necessary for the die-off of 90% of bacteria) varied from a maximum 40 hours during nighttime to a minimum of 2 hours in daytime.

4.2.1   Occurrence of Viruses in Marine Waters

The contamination of seawater by viruses occurs mainly through the disposal of sewage or sewage effluents into estuarine waters, offshore disposal via sewage outfalls, and rivers contaminated with domestic sewage.

A survey carried out in the Mediterranean sea by the Environmental Health Laboratory of Hebrew University, Israel, led to the detection of several types of enteroviruses (poliovirus, echovirus, coxsackievirus) as far as 1500 m from the discharge point. Coliform monitoring showed that these bacteria were less stable than viruses in seawater.

Another virus survey was carried out along the Houston ship channel and in Galveston Bay, Texas. This study showed that sewage treatment plants may

TABLE 4.3.   Enteric Viruses Recovered at Collecting Stations in Houston Ship Channel and Galveston Bay[a]

Station	Trial	Vol. (gal)	Enteric Virus Isolates						PFU/gal
			P1	P3	E1	E7	CB5	NT	
HSC-1[b]	1	85	17	0	8	17	25	17	0.9
	2	75	0	0	0	17	0	0	0.2
	3	131	25	8	0	133	0	217	2.9
	4	85	150	8	0	50	0	33	2.8
	5	35	17	0	0	17	0	0	0.9
	6	50	117	17	0	25	0	8	3.3
			326	33	8	259	25	275	
HSC-2	1	19	0	0	0	8	0	0	0.4
	2	26	25	0	0	17	0	0	1.6
	3	50	67	0	0	58	0	0	2.5
	4	45	17	0	0	0	0	0	0.3
	5	52	8	0	0	8	0	0	0.3
	6	78	8	8	0	17	0	0	0.4
			125	8	0	108	0	0	
GB-1	1–6	25–105	0	0	0	0	0	0	0

[a] From T. G. Metcalf, C. Wallis and J. L. Melnick (1974), in: *Virus Survival in Water and Wastewater*, J. F. Malina, Jr. and B. P. Sagik, Eds., The University of Austin, Texas.

[b] Total PFU's for HSC-1 station calculated from 277 actual isolations. Totals for HSC-2 station calculated from 73 actual isolations.

HSC-1   station located downstream 0.5 mile from WTP #2, and 4 miles from WTP #1.

HSC-2   station located downstream 4 miles from HSC-1 station.

GB-1   station located in Galveston Bay, 16 miles downstream from HSC-2 station.

Courtesy of Center for Research in Water Resources, The University of Texas at Austin.

TABLE 4.4.   Factors That May Affect the Survival of Enteric Viruses in Seawater

Parameters	Observation
A. PHYSICAL–CHEMICAL FACTORS	
Temperature	May be the decisive factor influencing virus inactivation; the loss of virus increases with temperature
Solar radiation	Probable effect at the surface of the water; turbidity protects viruses from effect of light
Adsorption to particulates and sedimentation	Particulates protect viruses and lead to their settling into marine sediments
Chemical composition of seawater	Salinity does not affect virus survival significantly; the effect of heavy metals is unknown
B. BIOLOGICAL FACTORS	
Biological inactivation (predation, enzymatic degradation)	Non-sterile seawater is generally more virucidal than autoclaved or filter sterilized seawater; marine bacteria with virucidal effect have been isolated
Type of virus	Seawater antiviral activity varies with the type of virus

discharge from 70×10^6 to 1746×10^6 viruses per day into these waters. Examination of the Houston ship channel for viruses (Table 4.3) led to the detection of enteroviruses as far as 8 miles (=13 km) from the contamination source. More recently viruses were found in Galveston Bay on many occasions. These examples show that one has to recognize that marine waters may be contaminated by viruses and that these pathogens appear to survive for longer periods than bacteria. These findings prompted worldwide research on factors controlling the inactivation of viruses in marine waters.

4.2.2   Virus Survival in Marine Waters

The phenomenon describing the virucidal power of seawater was first reported in France in 1961. Since then, laboratory studies have shown that virus survival is primarily controlled by physical, chemical and biological factors.

However, "the virucidal agent" remains yet to be identified. Some of these laboratory experiments were confirmed by "in situ" studies. We will now examine in more detail the various factors contributing to the inactivation or persistence of viral pathogens in seawater. These factors are summarized in Table 4.4.

Physical–Chemical Factors

Temperature. It was repeatedly demonstrated that temperature is the most decisive factor controlling virus in seawater and in other coastal waters as well. Table 4.5 illustrates the importance of temperature in enteric virus persistence in seawater. It is clearly indicated that viral inactivation is higher as the water temperature increases. The time necessary for 99.9% decline in virus titers is approximately 40–90 days at 3–5°C, 2.5–9 days at 22–25°C, and less than 5 days at 37°C. A typical example of thermal inactivation in seawater is given in Figure 4.4. Temperature may also have an indirect influence on the fate of viruses in seawater. In fact it has been shown in laboratory experiments that

TABLE 4.5.   Inactivation of Enteric Viruses in Seawater

Location	Virus Type	Temperature (°C)	Loss of 3 Log_{10} (days)
Mediterranean Coast (Tel Aviv, Israel)	Poliovirus Type 1	22 ± 3	2.5
		15	9
Mediterranean Sea (Beirut, Lebanon)	Poliovirus Type 1	25	6–7
Red Sea (Israel)	Poliovirus Type 1	22 ± 3	6
Gulf Coast (Mobile Bay, Alabama)	Poliovirus Type 1	24 ± 2	5–6
Atlantic Ocean (Offshore from France)	Poliovirus Type 2	4	>36
		12	12
		22	9
Baltic and North Seas	Poliovirus Type 3	23	8 (loss of 2–4 logs)
Pacific Ocean (offshore from the Van Dam State Park, California)	Echovirus 6	3–5	>40
		22	2.5
Atlantic Ocean (offshore from New Hampshire)	Echovirus 6	5	90
		37	<5
	Coxsackievirus B3	24	30
Gulf Coast (USA)	Coxsackievirus B1	20–25	3
	Reovirus 1	20–25	4

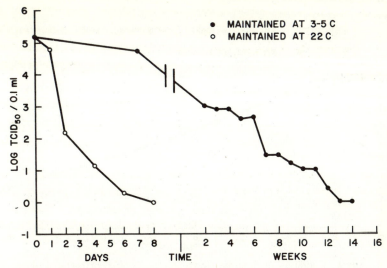

FIG. 4.4.   Inactivation rate of Echovirus 6 in seawater as a function of tempera-ture. From: W. D. Won and H. Ross (1973), *J. Environ. Eng. Div.* **99**:205–211. Courtesy of American Society of Civil Engineers.

the season of water sampling was of importance in this phenomenon. The bio-logical and chemical components in natural waters vary seasonally, and this may in turn influence the decline of viruses.

Solar Radiation. We have already emphasized the significance of sunlight in the die-off of *E. coli* in seawater (Figure 4.3). It remains yet to be demonstrated whether this factor plays any significant role in virus decline in coastal waters.

Adsorption to Particulates and Sedimentation. Organic and inorganic particulate matter, naturally present in sewage and seawater, exerts some influence on the fate of enteric bacteria and viruses. Viruses readily attach to these particulates and settle to the sediments where they accumulate and reach high concentrations. This important topic is covered in more detail in Sections 4.4 and 4.5. Studies dealing with bacterial phages have shown that these microorganisms are protected from inactivation in seawater by organic colloids, montmorillonite (Figure 4.5), or kaolinite (Figure 4.6). It has been hypothesized that clay minerals protect viruses by adsorption and hence inacti-vation of toxins or extracellular enzymes that might be produced by antagonistic marine microorganisms. With regards to plant viruses, it has long been observed that they are protected from ribonuclease activity by bentonite, a three-layer clay.

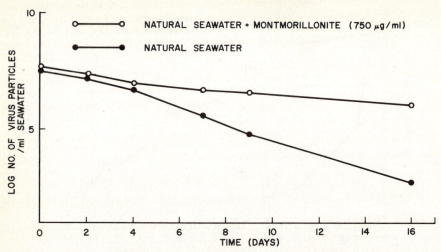

FIG. 4.5.   Effect of addition of montmorillonite on the survival of bacterial phage T7 in seawater. Adapted from G. Bitton and R. Mitchell (1974), *Water Res.* **8:**227–229.

Chemical Composition of Seawater. The salinity of seawater generally ranges from 3 to 3.8%. This level is actually lower in estuarine waters due to input of freshwater. Although salt, at a concentration of 1% (w/w) may impart some virucidal properties to distilled water, it has never been implicated in virus decline in seawater. Moreover, viruses may undergo aggregation as a

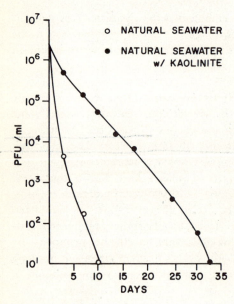

FIG. 4.6.   Effect of addition of kaolinite on the survival of bacterial phage T2 in seawater. Adapted from C. P. Gerba and G. E. Schaiberger (1975), *J. Water Poll. Control Fed.* **47:**93–103.

result of the high salt content of seawater, but this was shown to play a minor role in virus disappearance.

In the vicinity of ocean outfalls there is a constant mixing of sewage effluents with seawater, and it is tempting to think that proteinaceous materials present in wastewater may have a protective effect on viruses. This was found to be true in some studies, but in some others no appreciable effect was observed. It is, however, probable that organic particulates may actually protect viruses in the marine environment.

It has also been suggested that heavy metals could inactivate viruses in seawater, but this has yet to be proven.

Biological Factors

Any nonmarine microorganism . (bacteria, algae, fungi, virus) entering seawater may be subject to inactivation. Physical and chemical stresses imposed on these microorganisms predispose them to biological action (predation or enzymatic degradation). For example, it is known that in coastal waters enteric bacteria are readily destroyed by marine bacteria (pseudomonads), amoebic protozoa (Vexillifera), and by *Bdellovibrio bacteriovorus,* a tiny (0.3 µm) lytic bacterium.

Since the first studies undertaken by Plissier and Therre in Nice, France, it has been repeatedly shown that a heat-labile substance was involved in virus decline in seawater: viruses survive better in autoclaved than in raw seawater

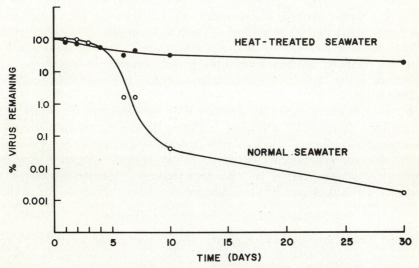

FIG. 4.7. Inactivation of poliovirus in normal and heat-treated seawater at 15°C. From H. I. Shuval et al. (1971), *J. San. Eng. Div.* **97**:587–600 Courtesy of American Society of Civil Engineers.

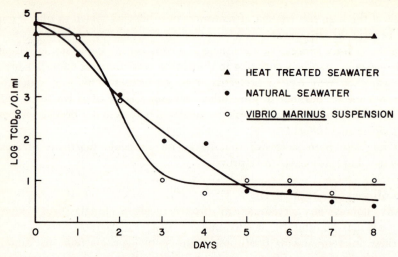

FIG. 4.8. Effect of *Vibrio marinus* on viral inactivation in seawater. From: Magnusson et al. (1967), *Acta Pathol. Microbiol. Scand.* **71**:274–280. Courtesy of *Acta Path. Microbiol. Scand.*

(Figure 4.7). This antiviral activity has been called VIC (virus inactivating capacity), MAVA (marine antiviral activity), LOT (loss of titer), or LOI (loss of infectivity). Attempts have been made to pinpoint the source of this antiviral activity. Marine bacteria have been implicated, and a relationship was demonstrated between their level in seawater and antiviral activity. One of these lytic marine bacteria was identified as *Vibrio marinus*. Figure 4.8 shows the degree of inactivation of poliovirus type 3 in the presence of this bacterium. The antiviral activity was lost after subculturing the bacteria at room temperature, but remained unaffected when the growth temperature was lower. It is thus possible that the antiviral action is associated with the metabolic activity of marine bacteria.

Inactivation in seawater may also vary from one virus type to another. It is rather difficult to make a general statement on the extent of inactivation in seawater according to virus identity because survival data have been generated under different experimental conditions (assay methods, origin of seawater, agitation of samples). However, it has been reported in some studies that, with respect to survival, viruses behave as follows:

coxsackievirus > echovirus > poliovirus

Figure 4.9 illustrates the stability of these viruses in estuarine water.

In Situ Survival Studies

It has been observed that viruses die faster under field than under laboratory conditions. The mechanism behind this phenomen is not clear.

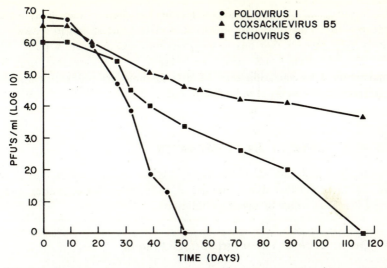

FIG. 4.9. Survival of enteroviruses in natural estuarine water during winter months. From S. Lo et al. (1976), *Appl. Environ. Microbiol.* **32**:245–249. Courtesy of American Society for Microbiology.

Dialysis bags and flow-through systems are generally used for assessing virus survival under *in situ* conditions. Table 4.6 shows the survival of three virus types under winter and summer conditions. Thus it can be seen that viral stability depends on virus identity and season (temperature) of the year.

We have seen that the virucidal activity of seawater is a widespread phenomenon, which is controlled by physical, chemical, and biological parameters. These factors do not completely remove the virus threat from

TABLE 4.6.   *In Situ* Survival of Enteroviruses in Ocean Water[a,b]

	Time (days) to Cause Log_{10} Virus Reduction in Virus Titer	
Virus Type	Summer $T = 21–26°C$	Winter $T = 4–16°C$
Poliovirus 1	20	40
Echovirus 6	22	55
Coxsackievirus B5	35	>80

[a] Adapted from S. Lo, J. Giblert, and F. Hetrick (1976), *Appl. Environ. Microbiol.* **32**:245–249.

[b] Virus suspensions were placed in dialysis bags.

marine waters, since the pathogens are sometimes found miles away from the discharge point. This obviously means that virus inactivation in seawater is a relatively slow process that has to be taken into account from a public health standpoint. Moreover, the virus inactivation rate is variable and hence unpredictable. This variation may be due to some yet unidentified seawater component(s) that remain(s) to be discovered.

4.3 VIRUS SURVIVAL IN FRESHWATER

The occurrence and survival of viruses in surface waters (rivers, lakes) are of public health significance due to the use of these waters for recreation and as a source of drinking water by urban populations.

4.3.1 Occurrence of Viruses in Surface Waters

Virus may enter lakes and rivers following pollution by septic tank or sewage treatment plant effluents. The problem is particularly acute in rivers that receive inadequately or nontreated wastewater originating from heavily populated areas. The virus load in wastewater tends to increase with the size of the community. This is illustrated in Table 4.7 which shows the results of a virus survey of the wastewater from four French towns during the period of 1961 to 1963. Although the detection method (gauze pad) used was not very efficient, viruses were present in all the river water samples. The percentage of samples positive for virus varied from 5.3 to 18.1. The predominant enteric viruses in wastewater or river water belong to the coxsackie B group (Table 4.8), which was also found to be the most resistant group in laboratory experiments. The

TABLE 4.7. Geographical Distribution of Viruses in Wastewater and River Water in France (1961–1963)[a]

Location	Population	% Positive Samples	
		Wastewater	River Water
Blainville	4,321	14.3	5.3
Baccarat	6,164	29.1	18.1
Luneville	24,463	44.8	6.4
Nancy	133,532	46.3	10.5

[a] Adapted from J. M. Foliguet, L. Schwartzbrod, and O. G. Gaudin (1966), *Bull. WHO* **35**:737–749.

TABLE 4.8.   Virus Types in Wastewater and River Water[a]

	% Positive Samples	
Virus Type	Wastewater	River Water
Poliovirus (mainly polio 1)	16.3	11.5
Coxsackievirus (mainly coxsackie B)	44.5	46.1
Echovirus (mainly type 1 and 7)	21.8	11.5
Reovirus	3.6	—
Unidentified virus	13.6	30.5

[a] Survey carried out in France (Meurthe and Moselle region) from 1961 to 1963. Data adapted from J. M. Foliguet, L. Schwartzbrod, and O. G. Gaudin (1966), *Bull. WHO* **35**:737–749.

phenomenon of viral pollution of rivers has been documented all around the globe and must be taken into account when one considers river water for potable or recreational use. In Table 4.9 one can see that 9 to 63% of river water in many parts of the world may be positive for viruses.

4.3.2   Virus Survival in Freshwater

Virus survival in freshwater generally has not been as well documented as in seawater. From a survey of the literature it appears that *temperature is the most significant factor controlling virus survival in lake and river waters.* Table 4.10 shows a summary of survival data at 3–6°C and 18–27°C temperature ranges. Enteroviruses display a 3 log reduction in 7 to 67 days at 3–6°C, and in 3 to 16 days at the 18–27°C temperature range. It also appears that coxsackieviruses are less resistant in freshwater environment than polio- or echoviruses. This is obviously a rather general picture of virus survival. Indeed, a closer examination of the data generated during the last decade shows, for example, that coxsackie B are less subject to inactivation than coxsackie A viruses. A study of virus inactivation in Lake Wingra, Wisconsin showed that coxsackievirus A9 is less stable than poliovirus 1 (Figure 4.10). Moreover, coxsackievirus B1 is more stable than poliovirus 1 in Rio Grande water (Figure 4.11). Other enteric viruses, such as reoviruses or adenoviruses, have not yet been examined for their stability in river or lake water.

TABLE 4.9. Viral Pollution in Rivers Around the Globe[a]

River	% Samples Positive for Viruses
Seine River	24
River Lee	33
River Thames	46–56
River Rhur	26
2 Swiss rivers emptying into Lake Geneva	38–63
Jordan River	9
Russian River (near Moscow)	34
Illinois River	27

[a] Data adapted from E. W. Akin, W. H. Benton, and W. F. Hill, Jr. (1973), in: Virus and Water Quality: Occurrence and Control, 13th Water Qual. Conf., University of Illinois at Urbana–Champaign.

As was found in seawater, heat-labile inactivating factors seem to be responsible for some of the viral inactivation, and these pathogens are generally more stable in autoclaved than in raw or filter-sterilized water.

A recent study has, however, demonstrated the importance of sunlight in virus inactivation in river water.

As previously discussed for the marine environment, it is not possible at the present time to pinpoint the factor(s) inactivating viruses in freshwater environ-

TABLE 4.10. Survival of Enteroviruses in Freshwater Environment

Virus Type	Time (Days) Necessary for 3 log Reduction in Virus Survival	
	3–6°C	18–27°C
Poliovirus	19–67	4–16
Coxsackievirus	7–10	3–6
Echovirus	15–60	5–16

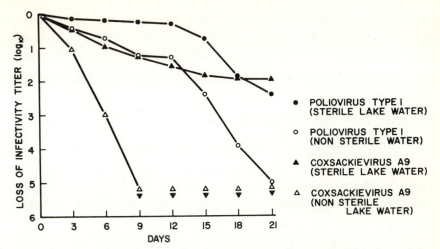

FIG. 4.10. Inactivation of enteroviruses in lake water (Lake Wingra, Wis.). From: J. E. Herrmann et al. (1974), *Appl. Microbiol.* **28**:895–896. Courtesy of American Society for Microbiology.

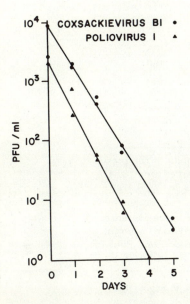

FIG. 4.11. Inactivation of poliovirus 1 and cox-sackievirus B1 in Rio Grande water at 23–27°C. From: R. T. O'Brien and J. S. Newman (1977), *Appl. Environ. Microbiol.* **33**:334–340. Courtesy of American Society for Microbiology.

ment. However, whatever the factor involved it has been shown that following exposure to river water the capsid integrity is sufficiently altered to bring about the degradation of the nucleic acid core, probably by enzymes called nucleases.

4.4 ASSOCIATION OF VIRUSES WITH PARTICULATES IN THE AQUATIC ENVIRONMENT

4.4.1 Adsorption of Viruses to Suspended Solids

Chapter 6 shows that viruses are routinely associated with wastewater suspended solids and the implications of this attachment process on the extent of virus removal by wastewater treatment plants. Viruses display also an affinity

TABLE 4.11. Infectivity of Viruses in the Adsorbed State

Virus	% Infectious Virus
Virus adsorbed to bentonite clay[a]	
T2	82
T7	76
f2	3
Poliovirus 1	93
Virus adsorbed to magnetite[b]	
T2	75
MS2	100
Poliovirus 1	100
Natural solids from lakes[c]	
Encephalomyocarditis virus (EMC)	87–138

[a] Viruses were adsorbed to bentonite in the presence of 0.01 M $CaCl_2$. Data adapted from B. E. Moore, B. P. Sagik, and J. F. Malina, Jr. (1975), *Water Res.* **9**:197–203.

[b] Virus was adsorbed to magnetite in the presence of 1610 ppm of $CaCl_2$. Data adapted from G. Bitton, G. E. Gifford, and O. C. Pancorbo (1976), Publication No. 40, Florida Water Resources Res. Center.

[c] Adapted from S. A. Schaub and B. P. Sagik (1975), *Appl. Microbiol.* **30**:212–222.

TABLE 4.12.   Accumulation of Indicator Bacteria in Freshwater and Marine Sediments

	Marine Sediment[a] (MPN/100 ml)		Freshwater Sediment[b] (CFU/cm³ or cm²)	
	Total Coliform	Fecal Coliform	Total Coliform	Fecal Coliform
Overlying Water				
Station #1	6,886	2,382	350	5.9
2	5,320	1,528	790	45
3	64	19		
4	92	10		
Sediment				
Station #1	382,143	9,731	71,000	3,400
2	192,857	16,806	67,000	2,700
3	16,791	152		
4	14,279	151		

[a] Sediments of canal communities along the Texas coast. Data adapted from S. M. Goyal, C. P. Gerba, and J. L. Melnick, 1977, *Appl. Environ. Microbiol.* **34**:139–149.

[b] Sediments from Shetuket River in Connecticut. Adapted from E. A. Matson, S. G. Hornor, and J. D. Buck (1978). *J. Water Poll. Control Fed.* **50**:13–19.

for suspended solids in the aquatic environment. These solids may be silts, clay minerals, cell debris, or particulate organic matter in general. Marine and river silts have been shown to adsorb substantial amounts of enteric viruses. However, because of their high exchange capacity and large surface area clay minerals have received most of the attention in the last few years. The main properties of clay minerals are reviewed in Chapter 9. In general terms, the adsorption of viruses to clay minerals depends on the type of clay, concentration and valency of cations in the suspending medium, pH, and the concentration of competitive organic materials. The adsorption process is rapid and may sometimes take only a few minutes.

The association of viruses to solids is of great importance from a public health standpoint:

- Solid-associated viruses must be taken into account in any virus detection method.
- Solid-associated viruses survive longer in natural water.
- Solid-associated viruses may settle and accumulate in the sediments.

4.4.2 Infectivity of Solid-Associated Viruses

It has been shown, in recent years, that solid-bound viruses are generally as infective as "free" virions. Table 4.11 shows the infectivity of bacterial phages and animal viruses associated with clay minerals, iron oxides, or natural solids present in lake water. It is apparent that most of the viruses are infective in the adsorbed state. The only exception was f2, an RNA phage, and the mechanism behind this phenomenon is not well understood.

In order to initiate infection, solid-associated viruses must attach to the surface receptors on host cells. This is probably achieved through the dissociation of the virions from solids prior to adsorption to host cells.

4.4.3 Role of Solids in Virus Survival

We have previously seen that solids (montmorillonite, kaolinite) prolong the survival of viruses in seawater. In general, clay minerals protect viruses from biological (enzyme action) and physicochemical (temperature, pH, ultraviolet

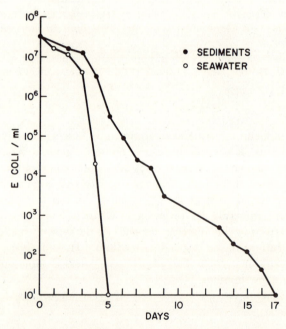

FIG. 4.12. Fate of *E. coli* in a marine sediment and in seawater. From C. P. Gerba and J. S. McLeod (1976), *Appl. Environ. Microbiol.* **32**:114–120. Courtesy of American Society for Microbiology.

TABLE 4.13. Accumulation of Enteric Viruses in Estuarine and Marine Sediments

Type of Sediment and Station Number	Virus Concentration (PFU/100 l)	
	Water	Sediment
Marine sediment[a]		
1 (Miami)	0.5	2160
2 (Miami Beach)	7.3	9830
Estuarine sediment[b]		
1	160	2000
2	90	2080

[a] Data adapted from C. P. Gerba, E. M. Smith, G. E. Schaiberger, and T. M. Edmond (1978), ASM Meeting, Las Vegas, Nevada. At station 1 (Miami) there was discharge of a chlorinated secondary effluent whereas at station 2 (Miami Beach), the sewage was untreated.

[b] Adapted from C. P. Gerba, S. M. Goyal, E. M. Smith, and J. L. Melnick (1977). *Marine Poll. Bull.* 8:279–282. Water and sediments were sampled in canal communities along the Texas coast.

and visible light) inactivation. This protective role has been widely demonstrated for bacterial survival in water, sediments, and soils. Furthermore, Chapter 8 examines the role of these solids in protecting viruses during the disinfection process.

4.5 THE FATE OF VIRUSES IN SEDIMENTS

Sediments are significant components of aquatic ecosystems and play a major role in biogeochemical cycles in general and microbial ecology in particular. Microorganisms, associated with organic and inorganic particulates, tend to settle into the bottom sediments where they reach high concentrations. This phenomenon is rather general and concerns bacteria as well as viruses. It is known that, at least in estuarine environments, coliform bacteria settle into sediment once a critical concentration of electrolyte is reached. It has been well established that there is a buildup of indicator bacteria in sediments around ocean outfalls where nontreated wastewater effluents are discharged into the ocean. Table 4.12 shows that, whether in marine or freshwater environments, there are greater numbers of indicator bacteria in sediments than in the overlying water. Enteric pathogens such as salmonellae are also isolated with greater frequency from sediments than from the water column. The accumulation of enteric pathogens in sediments is due to their longer survival in that particular environment. Figure 4.12 illustrates the higher survival of *E. coli* (an indicator of fecal pollution) in sediment than in seawater. Nutrients present in sediments may support the growth of this bacterium. Sediment particulates protect indicator bacteria and probably enteric pathogens from phage attack, and from predation by *Bdellovibrio* or enzyme-producing myxobacteria. Thus particulates in sediment may interfere with predator–prey or host–parasite relationships and this may explain the higher survival of enteric bacteria in sediments.

The question of virus occurrence and survival in sediments was addressed only after improved detection methods (for details on virus detection in sediments, see Chapter 5) were at hand. At the present time the efficiency of these methods is approximately 50%. This has allowed the monitoring of enteric viruses in sediments, particularly in those around ocean outfalls. Enteric viruses tend to follow the same pattern as enteric bacteria: they are present in higher numbers in sediments than in the overlying water. This pattern is illustrated in Table 4.13, which demonstrates the accumulation of viruses in marine and estuarine sediments. As demonstrated for bacteria, it is probable that viruses accumulate in the upper layer of sediments.

Viruses survive longer in sediments than in the overlying water. A study of virus survival in a sandy marine sediment has shown that the inactivation of poliovirus type 1 was 4.5-fold faster in seawater than in the sandy material. Other enteroviruses have been examined for their survival in marine sediments.

TABLE 4.14.   Survival of Selected Enteroviruses in Estuarine Water and Sediment[a]

Virus	Maximum Length (Days) of Virus Survival	
	Water	Sediment
Poliovirus 1	10	14
Coxsackievirus B3	4	18
Coxsackievirus A9	2	4
Echovirus 1	7	>18

[a] Adapted from E. M. Smith, C. P. Gerba, and J. L. Melnick (1978), *Appl. Environ. Microbiol.* **35**:685–689.

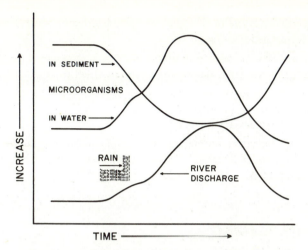

FIG. 4.13. Model of the relative changes in numbers of microorganisms in sediment and water during changing river discharge rates. From E. A. Matson et al. (1978), *J. Water Poll. Control Fed.* **50:**13–19. Courtesy of Water Pollution Control Federation.

They display a similar behavior, that is, a longer survival in the bottom sediments (Table 4.14). The mechanism of this protective effect is not clear yet. Therefore sediments may serve as reservoirs for viruses as well as for enteric bacteria. One may wonder about the public health significance of this phenomenon. There should not be any threat to public health as long as viruses remain absorbed to the sediments. However, resuspension of the upper layers of sediments would increase the virus concentration in the water column. Sediment resuspension is particularly significant in shallow waters, and it is accomplished through motor boat activity, swimming, currents, changes in river discharge rates, activity of benthic macroorganisms, dredging and changes in water quality (e.g., decrease in salinity in estuarine environment). With regard to river sediments, a model has been recently proposed to explain the changes in numbers of microorganisms in sediment and water during changes in river discharge rates (Figure 4.13). This model shows that following rains the river discharge increases and results in a resuspension of the upper layer of sediments with subsequent increase of microorganisms in the water column. This model is probably valid for enteric viruses.

4.6 SUMMARY

1. Physical, chemical, and biological factors control viral persistence in aquatic environments (marine- and freshwaters).

2. Temperature is the most significant physical factor controlling virus survival in natural waters.

3. Although marine microorganisms have been implicated in virus decline in seawater, the virucidal agent(s) remain(s) yet to be identified.

4. Viruses attach to suspended solids in aquatic systems. This association has implications in virus detection and survival. Most of the solid-associated viruses are infective.

5. Viral and bacterial pathogens settle into bottom sediments where they reach high concentrations. Sediments display a protective effect toward these pathogens.

4.7 FURTHER READING

Akin, E. W., W. H. Benton, and W. F. Hill, Jr. 1973. Enteric viruses in ground and surface waters: A review of their occurrence and survival. In: *Virus and Water Quality: Occurrence and Control*. 13th Water Quality Conference Univ. of Illinois at Urbana–Champaign.

Bitton, G. 1978. Survival of enteric viruses. In: *Water Pollution Microbiology, Vol.2*. R. Mitchell, ed. Wiley-Interscience, New York.

Foliguet, J. M., L. Schwartzbrod, and O. G. Gaudin. 1966. La Pollution virale des eaux usées, de surface et d'alimentation: Etude effectuée dans le département français de Meurthe-et-Moselle. *Bull. WHO* 35:737–749.

Mitchell, R., and C. Chamberlin. 1975. Factors influencing the survival of enteric microorganisms in the sea: An overview. In: *Symp. on Discharge of Sewage from Sea Outfalls*. A. L. H. Gameson, Ed. Pergamon Press, London.

five
Virus Detection in Water and Wastewater

Municipal wastewaters may harbor from 10^3 to 10^6 viruses per liter and this concentration may even be higher in some countries where socioeconomic conditions are low and where there may be a shortage of water that dilutes the sewage generated by urban populations. No virus concentration would be necessary if the wastewater contained 10^4 or more viruses per liter. After biological treatment the virus content of sewage is reduced by 1 or 2 orders of magnitude, and virus concentration becomes a necessity. This need is even higher when one contemplates the detection of viruses in comparatively highly treated water such as tapwater.

5.1 VIRUS SAMPLING

Unless the water is grossly polluted (sewage or wastewater effluent) one must take fairly large volumes of water, ranging from 10 to 100 gallons (38 to 380 liters) in order to detect small amounts of viruses. This is especially true for highly treated water such as drinking water. In that case the water sample volume could be as high as 1000 gallons (3780 liters).

In raw wastewater and, sometimes, in secondarily treated effluents, 1 to 4 liter samples are generally sufficient for the detection of viruses.

Whenever the water samples needed are larger, it is advisable to perform virus concentration in the field to avoid transportation problems.

5.2 SURVEY OF VIRUS CONCENTRATION METHODS

Various methods have been proposed for the concentration of low numbers of viruses from large volumes of water. Most of them have been tested under laboratory conditions and there is increasing concern over their performance under field situation. Most of these methods exploit some of the basic features of virus particles, that is, their surface properties, size, similarity to proteins, resistance to low temperatures, or their molecular weight.

In this section we will discuss the principles behind each concentration method. The advantages of the techniques as well as the problems encountered will also be considered. We will emphasize the membrane adsorption technique due to its popularity and adaptability to field conditions.

5.2.1 Gauze Pad Method (Swab Method)

This method consists of placing sanitary napkins in the water for a period of one or two days. Viruses, generally attached to solid particles in the water, are trapped within the swab matrix. Afterward the napkins are transported to the laboratory and eluted at pH 8 to 9. This method is generally suitable for wastewater or wastewater-treated effluents. It is a qualitative procedure which merely detects the presence or absence of viruses in water. The quantitation of viruses by this method has been attempted and a flow-through gauze sampler has been proposed (Figure 5.1). This sampler enables the measurement of the amount of water flowing through the pad. Unfortunately, the efficiency of virus

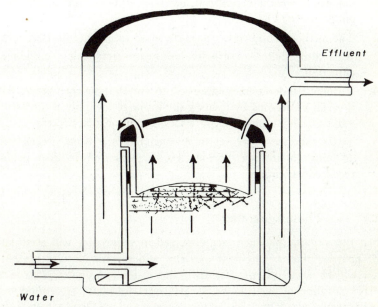

FIG. 5.1. Flow-through gauze sampler. From B. Fattal et al., (1974), in: *Virus Survival in Water and Wastewater Systems*, J. F. Malina, Jr. and B. P. Sagik, Eds., University of Texas, Austin. Courtesy of the Center For Research in Water Resources, University of Texas at Austin.

recovery is low. Although inexpensive and somewhat convenient, the gauze pad method is of limited use.

5.2.2   Hydroextraction

The test sample is placed in a dialysis bag which is surrounded by a hygroscopic agent, polyethyleneglycol (PEG). The polymer has a molecular weight equal to 4000 or 20,000. Water moves freely across the dialysis membrane, whereas viruses remain inside the bag. This procedure is generally completed within 18 to 24 hours and should be carried out at low temperature (4°C) to avoid virus inactivation.

The hydroextraction method has the advantage of being inexpensive and relatively simple with no capital cost involved. However, it has also many disadvantages:

1. The recovery is low and range from 3 to 30%. However, it was recently reported that recoveries of 90 to 100% are possible provided one uses small pore-sized dialysis membrane (25 Å average pore size), high molecular weight polymer (Carbowax 20,000), and incubation at low temperature (4°C).

2. The method is limited to relatively small volumes (100–1000 ml).

3. It is also limited to grossly polluted waters (e.g., secondary-treated effluents).

4. Viruses are sometimes adsorbed to the dialysis membrane. However, this can be avoided by treating the inside of the dialysis tube with a protein substance which interferes with the adsorption process.

5. Toxic substances present in sewage may inactivate viruses during the 24-hour incubation period. Hydroextraction is also useful as a reconcentration step.

5.2.3   Two-Phase Separation

When two organic polymers are dissolved in water they will eventually form two separate phases. Colloidal particles and macromolecules present in the water will be eventually partitioned in one of the two phases. This is the basis for concentration of particles by the polymer two-phase separation technique. The procedure was later used for the concentration of viruses from wastewater: two polymers (PEG and sodium dextran sulfate) and an appropriate salt (e.g., NaCl) are added to the test samples and the mixtures are incubated at 4°C for 18 to 24 hours. A two-phase system develops and viruses are concentrated in the dextran sulfate bottom phase (Figure 5.2). The partitioning of viruses is characterized by a partition coefficient K:

$$K = \frac{C_t}{C_b} \qquad (5.1)$$

where C_t = virus concentration in the top phase (PEG phase)
  C_b = virus concentration in the bottom phase (sodium dextran sulfate phase)

The concentration factor α is defined as:

$$\alpha = \frac{C_b}{C_i} \qquad (5.2)$$

where C_b = virus concentration in the bottom phase
  C_i = initial virus concentration in the test sample

From Equations 5.1 and 5.2, it is clear that low values of the partition coefficient K will result in a high concentration factor α.

The separation of viruses by this procedure greatly depends on pH conditions and on the ionic strength of the added salts. The cations have the ability to render the upper phase more negatively charged than the other, and consequently negatively charged viruses will be attracted to the more positively charged phase.

Although the polymer two-phase separation method is relatively simple and inexpensive, it is suitable only for small volumes (0.2 to 2 liters) of grossly polluted water. It cannot be used for the recovery of a low amount of viruses from large volumes of water. It has been reported that the efficiency of this method lies between 5 and 100%, and that sodium dextran sulfate could inhibit some enteric viruses such as certain strains of coxsackie- and echoviruses.

5.2.4 Ultracentrifugation, Freeze-Concentration, Ultrafiltration, and Electrophoresis

Ultracentrifugation. Ultracentrifugation of viruses at speeds above 50,000 rpm is effective in concentrating viruses from dilute suspensions. Unfortunately, the cost of equipment is prohibitive. Even continuous-flow ultracentrifugation can be time-consuming and is not indicated for the concentration of viruses from large volumes of water.

Freeze-Concentration. Viruses are highly resistant to low temperatures, and freezing at −70°C is traditionally used to store viral suspensions.

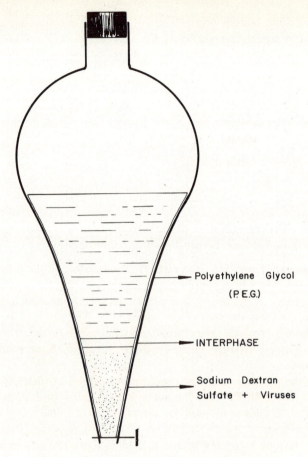

FIG. 5.2.   The polymer two-phase separation method.

In freeze-concentration the sample (1–5 liters) is cooled to −15°C under continuous stirring to avoid their being trapped in ice crystals. Virus, along with other colloidal particles and salts, are concentrated in the nonfrozen water that remains in the center of the container. This method, although simple and inexpensive, is not practical for the processing of large volumes of water.

Ultrafiltration. Ultrafiltration is the process by which water is driven through a membrane by applying pressure and, consequently, viruses and macromolecules are retained because of the small pore size of the ultrafilter. It separates the sample into two components:

- *Retentate.* Contains viruses and other macromolecules.

• *Filtrate.* Contains smaller molecules that pass through the filter.

This process suffers mainly from clogging of the membranes and this can be partially solved by maintaining the particles in suspension with an agitation device. The development of cellulosic hollow-fiber membranes has allowed the processing of relatively large volumes of water. In *tangential flow ultrafiltration,* the water flows over the surface of the membrane. Figure 5.3 shows an ultrafiltration device known as the Pellicon cassette system (Millipore Corp., Bedford, Mass.). The volume of processed water may be enhanced by increasing the surface area of the membrane, from 1 to 50 ft².

Ultrafiltration is now essentially used as a reconcentration step in virus detection.

In the laboratory the use of *soluble alginate filters* is based on a similar concept with the exception that the sodium alginate filter can be dissolved in a nontoxic solution (sodium citrate) and assayed directly. This procedure allows

FIG. 5.3. Tangential flow ultrafiltration device. (Courtesy of Millipore Corp., Bedford, Mass.)

a good recovery of viruses, is economical and relatively simple, but does not function well in turbid water due to clogging of the filter.

   Electrophoresis. Virus particles are negatively charged at neutral pH values and move toward the anode when placed in an electric field. This phenomenon has been used for concentrating phages and enteric viruses on dialysis membranes. Figure 5.4 is a diagram of a forced-flow electrophoresis cell where viruses adsorb to dialysis membranes and can be desorbed by simply reversing the polarity of the current.

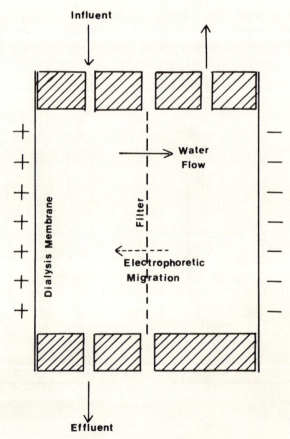

FIG. 5.4.   Forced flow electrophoresis cell. Adapted from M. Bier and F. C. Cooper. (1967). in: *Principles and Applications of Water Chemistry*, S. D. Faust and J. V. Hunter, Eds., Wiley, New York.

5.2.5   Methods Based on the Process of Adsorption to Surfaces

Adsorption of Viruses to Surfaces: Theoretical Considerations

Viruses are colloidal hydrophilic particles that can interact with solid surfaces. According to the pH of the water, they are negatively or positively charged and they exhibit an *isoelectric point* at which their charge is zero. At pH values found in natural waters viruses behave essentially as negatively charged colloids. In fact most surfaces in nature possess a net negative charge. We will now examine the mechanisms by which viruses are "attracted" to solid surfaces.

Brownian Motion. Viruses are not actively motile and must rely on Brownian motion and on the movement of the suspending fluid to approach a surface. Brownian motion is a random motion caused by the thermal motion of molecules following their collision with other molecules or with colloids.

The *Brownian displacement* (Δ) is given by the following equation:

$$\Delta = 2Dt \tag{5.3}$$

where D = diffusion coefficient
  t = time in seconds

The diffusion coefficient is inversely proportional to the size of the colloidal particles.

It has been shown experimentally that the rate at which some viruses come close to a surface can be predicted by the Brownian motion theory. For a virus within the size range of 10 to 100 nm, the Brownian displacement is approximately 0.12 to 0.4 mm/hour. For some viruses the encounter with a surface is probably diffusion-limited, and this may explain the delay in virus infection of some host cells.

Electrostatic Forces. The ionogenic groups present on both virus and solid surfaces govern the adsorption process. The electric charge on each surface will depend on the pH of the suspension. A virus is negatively charged above its isoelectric point and positively charged below that point. Electrostatic attraction will play a major role in the adsorption of viruses to positively charged surfaces like aluminum surfaces.

Electrical Double-Layer Phenomenon. The interaction between two surfaces is the result of repulsion and attraction forces.

The attraction forces act at short distances and are called *London–van der Waals forces.* Since the attraction energy remains constant, one has to reduce the repulsion forces in order to bring about an interaction between the two surfaces. This reduction is made possible by *increasing the concentration and valency of the suspending electrolyte.* The practical value of this concept will be considered in the next section.

Adsorption to Membrane Filters

Membrane filters are rapidly becoming an essential tool in virus detection methodology. They are increasingly used for the concentration of viruses from environmental waters, soils, sediments, and cell harvests.

Viruses are efficiently retained by membrane filters of various chemical structure and configuration. These membranes can be made of cellulose acetate, cellulose nitrate, fiberglass, epoxy–fiberglass–asbestos, and various other substances. They may be used as flat or cartridge-type filters. The latter offer a much larger surface area for the attachment of viruses (Table 5.1), and their use in the processing of large volumes of water will be discussed later. Membrane filters are able to retain viruses despite the fact that their pore size (generally 0.45 μm) is much larger than that of viruses (less than 30 nm for all the enteroviruses). This demonstrates that, except for solid-associated viruses, entrapment of virus particles plays a minor role in their removal by membrane filtration. Thus adsorption is undoubtedly the major process involved in viral retention by membrane filters.

Three possible mechanisms have been suggested to explain the adsorption of viruses to membrane surfaces:

1. *Hydrophobic Bonding.* The surface of viruses has nonpolar (hydrophobic) regions that may be bound to other nonpolar regions located on the surface of other viruses (this leads to virus aggregation) or membrane filters.

TABLE 5.1.   Surface Area of Flat Filters and Cartridge-type Filters[a]

Filter	Surface Area (cm²)
A. Cartridge filters	
Acropor (25.4 cm)	5600
Filterite (25.4 cm)	2800
Millipore (55.8 cm)	850
B. Flat filters	
293 mm in diameter	470
142 mm in diameter	97

[a] Adapted from S. R. Farrah et al. (1976), *Appl. Environ. Microbiol.* **31:**221–226.

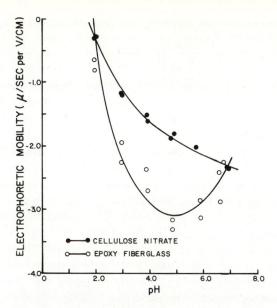

FIG. 5.5. Electrophoretic mobility vs pH of cellulose nitrate and epoxy fiberglass filter materials. Adapted from M. A. Kessick and R. A. Wagner (1978), *Water Res.* **12**:263–268.

2. ***Hydrogen Bonding.*** This occurs between polar groups on both virus and membrane filters.

3. ***Electrostatic Forces.*** Viruses and membrane filters are both negatively charged at pH values close to 7. Their interaction depends on the following factors:

   (a) ***Chemical Composition of the Filter.*** There are differences in the retention capacity of various membrane filters. For example, cellulose triacetate membranes are less efficient in viral retention than cellulose nitrate or epoxy fiberglass filters. Most of the membranes are usually negatively charged but some manufacturers have made available positively charged membrane filters which display high virus adsorption capacity at neutral pH.

   (b) ***pH.*** The net surface charge on the surface of both viruses and membrane filters can be controlled by changing the pH of the suspending medium. Viruses are generally positively charged at low pH values whereas membrane filters are negatively charged (negative charge due to the presence of ionized COO^- groups) at least between pH 2 and 8 (Figure 5.5). The surface charge of the filter material becomes less negative as the pH of the water decreases. "Zeta plus" filters (AMF, Cuno Division, Meriden, Conn.) are positively charged at pH values encountered in natural waters (pH 5–8). Within this pH range, the adsorption of viruses to these particular filters is high even in the absence of any added salt. Negatively charged filters (e.g., Filterite, Cox) necessitate the addition of cations for enhanced adsorption (Figure 5.6).

(c)   **Cations.** Virus adsorption to membrane filters increases with the concentration and valency of the cation present in the suspending medium. Cations decrease the negative charge on membrane filters and trivalent cations (e.g., Al) may even reverse it to the positive side. This phenomenon is illustrated in Figure 5.7. Hence in the process of virus concentration by membrane filtration, 0.0005 M aluminum chloride is as effective as 0.05 M magnesium chloride. This obviously is an important consideration when one contemplates the feasibility of the method under field conditions. Cations help decrease the repulsion forces between viruses and membrane surfaces. They may interact with specific groups and thus establish a "bridge" between the two surfaces. This is called *salt bridging*.

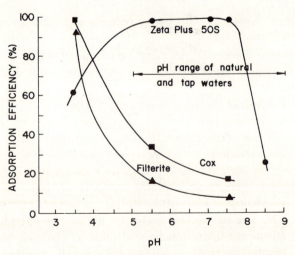

FIG. 5.6.   Poliovirus adsorption to filter materials as a function of pH. From B. L. Jones and M. D. Sobsey (1978), in: *98th Ann. Conf. Am. Water Works Assoc.*, Atlantic City, N.J. June 25–30, 1978. Courtesy of American Water Works Association.

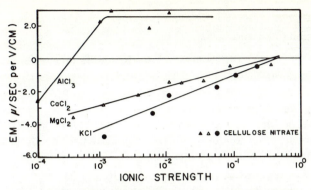

FIG. 5.7. Electrophoric mobility vs ionic strength for filter materials at pH 7.0. Adapted from M. A. Kessick and R. A. Wagner (1978), *Water Res.* **12:**263–268.

(d) *Flow Rate.* Viral retention to membrane filters decreases with increasing flow rates. Under appropriate conditions (i.e., low pH, presence of salts) viral adsorption is not reduced under high flow rates.

(e) *Organic Materials.* Organic substances (e.g., proteins) interfere with viruses for adsorption sites on membrane filters and other surfaces. These organics are soluble or colloidal materials collectively called *membrane coating components* (MCC). Humic substances are also known to coat membrane filters and thus reduce virus adsorption.

We have so far considered the various factors which may enhance or reduce the adsorption of viruses to membrane filters. Once a virus is retained on a membrane one must be able to elute (desorb) it from that surface in order to assay it in an appropriate host system. Proteinaceous fluids adjusted to high pH are generally used for that purpose. The most commonly used eluent is 0.05 *M* glycine buffer adjusted to pH = 11.5 with NaOH. Other eluents (beef extract, casein, nutrient broth, tryptose phosphate broth, arginine, lysine) are used at a lower pH (=9.0) which may be more suitable for certain viruses (adenoviruses, rotaviruses) that are rapidly inactivated when the pH exceeds 10.

The efficiency of the membrane filtration method depends on the following factors:

1. Selection of an appropriate membrane filter. For instance, nitrocellulose membranes are better adsorbents than those made of cellulose acetate.

2. Addition of salts to promote virus adsorption.

3. Adjustment of pH to low values (e.g., pH = 3.5).

4. Selection of an appropriate eluent.

Although membrane filtration is relatively simple and sensitive, it has the two following limitations:

- *Clogging.* Suspended solids present in natural waters may clog membrane filters. Since viruses are associated with this particulate matter, the proper use of prefilters is indicated. The prefilter must be included in the elution process.

- *Membrane coating components* (MCC). As stated earlier, these materials tend to interfere with the adsorption process.

Adsorption to Insoluble Polyelectrolytes

P.E. 60, an insoluble polyelectrolyte, has been widely used for the adsorption and subsequent concentration of viruses. It may be used in suspension or in layers sandwiched between two filter pads. Following adsorption the viruses are eluted with a proteinaceous solution at high pH. The mechanisms of adsorption is based on *hydrogen bonding* between the virus and chemical groups (e.g., carboxyl) on the polymer surface, and on electrostatic forces. The latter depend on pH and salt concentration.

Adsorption to Salt Precipitates

This method consists of adsorbing viruses on preformed flocs of aluminum hydroxide [$Al(OH)_3$], aluminum phosphate ($AlPO_4$) or calcium phosphate [$Ca_2(PO_4)_3$], and recovering the viruses by direct inoculation or dissolution of the flocs, or by elution of the viruses from the flocs. $Al(OH)_3$ precipitates display a good adsorption capacity toward viruses and allow the concentration of most viruses except reoviruses. The adsorption process is the result of electrostatic interaction between viruses and the positively charged surfaces of $Al(OH)_3$.

Although the recovery efficiency is good, this method is limited to small volumes of test water (up to 10 liters) and is essentially used as a reconcentration step.

Adsorption to Glass Powder

Viruses may be adsorbed to glass powder in the presence of 0.5 mM $AlCl_3$ at pH 3.5. When processing 50 liters of water at a flow rate of 80 liters/hour, 36 to more than 100% of input viruses have been recovered (Table 5.2). This method has yet to be evaluated for larger volumes of water.

TABLE 5.2.   Concentration of
Poliovirus by Adsorption to Glass
Powder[a]

PFU input (50 liters)	PFU in eluate (25 ml)	% Recovery
6500	3800	58
2700	980	36
75	77	103
75	68	90
18	7	39
10	8	80

[a] Adapted from L. Schwartzbrod and F. Lu-
cena-Gutierrez (1978), *Microbia.* **4**:55–68.

Adsorption to Iron Oxides

Iron oxides, such as hematite (Fe_2O_3) or magnetite (Fe_3O_4), are usually good adsorbents toward viruses present in water. Magnetite is generally more efficient than hematite. In a recent study it was found that the adsorption of poliovirus to magnetite was high at pH ranging between 5 and 9, and was not affected by the organics at levels generally present in wastewater effluents.

Three different approaches have been taken in the concentration of viruses by iron oxide:

1. The test water sample is passed through a *column packed with iron oxide*. Virus removal is high, but the column has a tendency to clog. The use of a prefilter to remove suspended solids will remove simultaneously a great number of viruses that are found attached to solids in water.

2. *"Sandwich" technique.* This approach consists of placing the iron oxide between two filter pads. However, this method suffers from the same problems as the column method.

3. *"Magnetic filtration" approach.* (Figure 5.8). In this method, the test water is seeded with magnetite at a concentration of 100 to 300 ppm. Under appropriate salt concentration, the virus adsorbs efficiently to the magnetite. The mixture is then poured through a filter packed with stainless steel wool and placed in a background magnetic field. The magnetite, along with the viruses, is efficiently retained in the filter and can be washed off by turning the magnetic field off. This method has the advantage of high flow rates and does not suffer from clogging as the two previous methods.

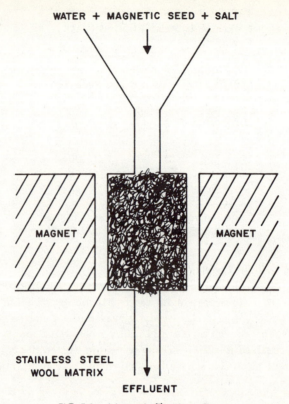

WATER + MAGNETIC SEED + SALT

MAGNET

MAGNET

STAINLESS STEEL
WOOL MATRIX

EFFLUENT

FIG. 5.8.   Magnetic filtration process.

Whatever the method adopted, viruses adsorbed to iron oxides can be eluted with a proteinaceous material (10% fetal calf serum or 1% casein) at pH 9.

Adsorption to Bentonite

Bentonite is an expandable three-layer clay mineral which is known for its high adsorptive capacity toward viruses and other materials. It offers viruses a huge surface area to adsorb on and the surface area can be as high as 800 m²/g of clay. Therefore, bentonite is sometimes used to concentrate viruses from water. The method consists of mixing the test sample with 100 to 1000 ppm of bentonite and adding $CaCl_2$ to a final concentration of 0.01 M. The mixtures are stirred for 30 minutes and allowed to settle overnight at 4°C. This technique can recover approximately 50% of the seeded virus.

A recent investigation has revealed that bentonite-virus mixtures can be trapped on flat fiberglass prefilters and that virus recovery may average approximately 70%. This procedure is faster than the settling method and it allows the processing of larger volumes of water.

5.3 FIELD MONITORING TECHNIQUES

5.3.1 The Problem of Solid-Associated Viruses in Water and Wastewater

Viruses are frequently associated with suspended solids in natural waters and wastewater. Viruses do not loose their infectivity following their adsorption to a solid particle. Most of the concentration techniques described so far fail to address the problem of solid-associated viruses and merely detect "free" virions. Efforts have been made to assess the percent of virus particles which are associated with natural solids. These enbedded viruses can be efficiently released by sonication or by some other drastic treatment. Table 5.3 shows that 68 to 90% of viruses are found in the solid portion of wastewater and secondarily treated effluents.

Therefore in any concentration technique one has to take into account these solid-bound viruses by adequately processing the solid portion.

5.3.2 General Methodology

Viruses may be present in tapwater, estuarine water, sewage effluents, groundwater or sanitary landfill leachates. There have been various attempts to concentrate viruses from large volumes (3.0 to more than 100 gallons) of these waters. In this section we will limit ourselves to the general concentration scheme under field situation. Membrane adsorption techniques have been so far

TABLE 5.3.   Solid-Associated Viruses in Wastewater[a]

	Liquid Portion		Solid Portion	
	No. PFU	%	No. PFU	%
Wastewater	45	19.2	189	80.8
	129	31.7	278	68.3
Wastewater effluent	4	16.7	20	83.3
	3	10.0	27	90.0

[a] Adapted from F. M. Wellings et al. (1977), in: *Wastewater Renovation and Re-use*. F. M. D'Itri, Ed. Marcel Dekker, New York.

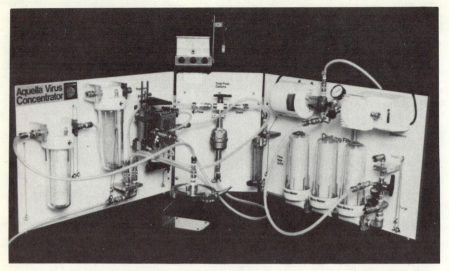

FIG. 5.9. "Wallis and Melnick" virus concentrator. (Courtesy of Carborundum Co., Niagara Falls, N.Y.)

used for this purpose. Two main types of membrane-adsorption methods are considered:

1. *Addition of celite* (diatomaceous earth) to water under appropriate conditions (pH and salt concentration). The mixture is subsequently passed through a flat membrane filter (293-mm diameter). The celite acts as a prefilter in this system.

2. The ***Wallis and Melnick Concentrator.*** This concentrator has been developed at the Baylor College of Medicine, Houston, Texas. It also concentrates viruses by adsorption to and elution from flat membrane filters or from cartridges (Figure 5.9). The various steps involved in this concentration scheme are shown in Figure 5.10. This concentrator was chosen as an example to illustrate the general methodology involved in virus concentration. A new version of the concentrator was proposed recently by EPA researchers.

The steps are as follows:

1. *Clarification or Prefiltration.* This step helps avoid the clogging problem encountered with some turbid waters. However, one has to process the prefilter to recover the solid-associated viruses. It is possible to process tap water or groundwater without any clarification step.

2. *Concentration.* It involves the adsorption of viruses to membrane filters at suitable pH (pH 3.5) and salt conditions (0.05 M $MgCl_2$ or 0.0005 M $AlCl_3$). Pleated Filterite cartridge filters have been shown to be very resistant to clogging and to enable the processing of water at flow rates of 10 gallons (37.8 liters)/minute instead of 1 to 2.5 gallons/minute attained with flat filters. Other advantages of the pleated membranes are shown in Table 5.4

3. *Elution.* 0.05 M glycine buffer at pH 11.5 is presently the most commonly used eluent in virus detection methodology. Concern over high pH-sensitive viruses (adenoviruses, reoviruses, rotaviruses) has led to a search for other eluents that function at lower pH values. Beef extract, casein, tryptose phosphate broth are examples of eluents which operate at pH = 9.0.

4. *Reconcentration.* A reconcentration step is necessary when large volumes of environmental waters are processed. This step reduces the volume of eluates to quantities that can be economically assayed on tissue cultures. Although many reconcentration methods have been proposed, only a few of them have been used by environmental virologists. These methods include:

   (a) *Membrane filtration.* The eluate is readsorbed to small membrane filters following pH adjustment and salt addition. However, humic substances and other organic materials may clog the filters and interfere with viral adsorption. Although they

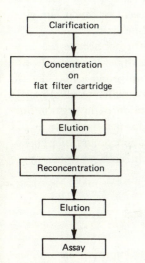

FIG. 5.10. Concentration scheme for the recovery of viruses from 50 gal (189 liters) or more of water.

TABLE 5.4.   Advantages of a Virus Concentrator Using Pleated Membrane Filters[a]

---

1.   The filter material used in their construction is the most resistant to clogging of all membrane filters tested

2.   Can be reused after autoclaving without loss of concentration efficiency

3.   Highest operating flow rates (up to 37.8 liters/minute) of any virus concentrator system

4.   Filter material can be bent or folded without damage

5.   Has been tested and found effective with a variety of waters with widely varying quality

6.   Low weight and size requirements (i.e., does not require bulky or heavy holders as in the case of flat disc filters)

7.   Can withstand operating differential pressures as high as 100 lb/in.^{2}

8.   Pleated filters easily fit into plastic housing described in previous models of the virus concentrator. This allows visual inspection of the filters at all times

---

[a] From C. P. Gerba et al. (1978), (unpublished).

can be removed by activated carbon or ion exchange treatment, it is advisable to consider some other reconcentration method when filter clogging occurs.

(b)   *Aluminum hydroxide flocculation* and *ferric chloride precipitation.* These methods have been successful in viral reconcentration from drinking and estuarine waters.

(c)   *Hydroextraction.* This method is described in Section 5.2.2.

(d)   *Organic flocculation.* The method consists of lowering the pH of a protein solution, acting as eluent, to produce flocculation, and then resuspending the sediment in a buffer at alkaline pH (pH = 9.0). Two eluents have been considered in organic flocculation: Beef extract and casein. The former flocculate at pH 3.5; whereas for the latter, the flocculation occurs at pH 4.5. The final recovery of poliovirus with both eluents ranges between 50 and 60% (Table 5.5).

5.3.3   Concentration of Viruses from Environmental Waters

Virus Concentration from Tapwater

Virus concentration methodology has considerably improved in the last decade and a standard method has been included in the 14th edition of *Stand-*

TABLE 5.5.   Comparison of Beef Extract to Casein for the Recovery of
Viruses from Seawater[a,b]

Eluent	Total Number of Viruses Adsorbed to Membrane Filters (PFU)	Viruses Eluted from Membrane Filter. % of Adsorbed Virus	Organic Flocculation (Reconcentration) % Recovery	Overall Recovery Efficiency (%)
3% Beef extract	3.50×10^6	78.5	65.1	51.1
	2.59×10^6	91.9	66.8	61.4
0.5% Purified	5.30×10^6	82.4	66.4	54.7
casein	4.43×10^6	125	53.2	66.4

[a] From G. Bitton, B. N. Feldberg, and S. R. Farrah (1979), Water, Air, Soil Poll. **12**:187–195.

[b] Poliovirus 1, suspended in seawater, was adsorbed to a series (3 μm → 0.45 μm → 0.25 μm) of Filterite filters in the presence of 0.0005 M AlCl$_3$ at pH = 3.5. The filters were eluted with 3% beef extract (pH = 9.0) or 0.5% purified casein (pH = 10.0). The eluates were then concentrated by organic flocculation.

ards Methods for the Examination of Water and Wastewater. This method is based on the membrane-adsorption-elution concept, which enables the processing of a minimum of 100 liters of tapwater. The technique has been further improved and it is now possible to process 1000 liters of tapwater, using Filterite cartridges and lower concentrations of aluminum chloride (2×10^{-5} M

TABLE 5.6.   Concentration of Poliovirus from 1000 Liters of Tapwater with 0.00002 M Aluminum Chloride[a]

Trial	Final Sample Volume (ml)	Poliovirus Recovered		
		PFU added	PFU	%[b]
1	80	1.7×10^6	1.3×10^6	76
2	30	7.3×10^6	5.7×10^6	79
3	20	5.7×10^6	4.5×10^6	79
4	38	2.2×10^6	1.3×10^6	59
5	28	176	112	64

[a] From S. R. Farrah et al. (1978), *Appl. Environ. Microbiol.*, **35**:624–626.

[b] Mean recovery = 71%.

TABLE 5.7.   Concentration of Poliovirus Type 1 from 50 Gallons
(189 Liters) of Tapwater by the Nonfat Dry Milk (NFDM) Technique[a]

Total Virus Input PFU	% Elution From Filters	Organic Flocculation % Recovery	Final Volume (ml)	Percent Overall Recovery
High Virus Input				
7.12×10^7	72	110	80	79
7.17×10^7	118	68	55	81
Low Virus Input[b]				
641	—	—	28	72
569	—	—	30	78

[a] From G. Bitton, B. N. Feldberg and S. R. Farrah (1979), Water, Air, Soil Poll. **12:**187–195. Poliovirus type 1, suspended in 50 gal (189 liters) of dechlorinated tapwater, was adsorbed to a 0.25 μm or 0.45 μm Filterite cartridge filter in the presence of 0.0005 M AlCl$_3$ at pH = 3.5. Viruses were eluted from the filters using 1% NFDM in 0.05 M glycine pH = 9.0. Filter eluates were brought to pH = 4.5 − 4.6 with 1 M glycine (pH = 2.0) and the resultant floc centrifuged at 4500 RPM for 4 minutes. The pellets were then resuspended in 10–67 ml Na$_2$HPO$_4$, pH = 9.0.

[b] Concentrates from low virus input trials were dialyzed for 18–24 hours against PBS at 4°C prior to direct assay on host cells.

AlCl$_3$) at ambient pH levels. The mean virus recovery is approximately 70% (Table 5.6).

These methods suffer, however, from the fact that some viruses (e.g., adenoviruses, rotaviruses) are inactivated by high pH. Viruses can be recovered at pH 9.0 with fetal calf serum, beef extract, casein or nonfat dry milk.

Enteroviruses can be efficiently adsorbed to talc-celite layers and then eluted with 10% fetal calf serum at pH = 9.0. The eluate is further concentrated by hydroextraction. This method may recover from 58 to 64% of poliovirus from 100 to 1000 liters of potable water. Poliovirus is efficiently recovered (72–81% efficiency) from 50 gallons (189 liters) of tapwater with 1% nonfat dry milk at pH 9.0. More details on this method are given in Table 5.7.

It was recently reported that 3% lysine at pH 9.0 could serve as an eluent for virus adsorbed to Filterite filters, while allowing virus adsorption to Zeta plus filters. This finding was used to develop a two-step concentration procedure for virus in tapwater (Table 5.8). The mean recovery of this method is 67%.

Poliovirus type 1 concentration from 100 liters of tapwater by *capillary ultrafiltration* resulted in 80% mean virus recovery (Table 5.9). In this method the ultrafiltration unit consists of capillaries (1.5 mm ID) with a molecular weight cut off of 50,000. The application of this method to polluted surface waters remains to be investigated.

Virus Concentration from Seawater

The need to monitor the occurrence of enteric viruses in ocean and estuary waters is now widely recognized. Early methods failed to recover significant percentages of input virus due to the turbidity of these waters. The introduction of pleated Epoxy–fiberglass cartridge filters (Filterite Corp.) has resulted in a significant improvement in virus recovery efficiency. This concentration method consists essentially of:

1. Adsorption of virus to an epoxy–fiberglass cartridge in presence of 0.0005 M to 0.0015 M AlCl$_3$ at pH = 3.5.

2. Elution with glycine buffer at pH = 11.5.

3. Reconcentration by Al or Fe flocculation followed by hydroextraction.

As previously discussed for tapwater, concern has been raised over the inactivation of some enteric virus by the high pH of glycine buffer used as eluent. It was thus proposed to use beef extract and casein, at pH 9, as eluents. The reconcentration step is done through organic flocculation. Overall virus recovery efficiency with both eluents can vary from 50 to 60% (Table 5.5).

TABLE 5.8.   Two-Step Concentration of Viruses from Tapwater Using Different Membrane Filters[a]

Trial	Virus in 4000 ml of Tapwater PFU	Virus Eluted from 47 mm Filterite Filters by 20 ml of 3% pH 9 Lysine PFU	Virus Eluted from 25 mm Zeta Plus Filters by 4 ml of 3% pH 9 Beef Extract	
			PFU	% of Added Virus
1	1.1×10^7	8.8×10^6	6.2×10^6	58
2	6.2×10^6	6.2×10^6	5.6×10^6	90
3	7.8×10^6	6.4×10^6	5.2×10^6	67
4	7.7×10^6	6.2×10^6	4.1×10^6	53
			Mean = 67	

[a] From S. R. Farrah and G. Bitton (1979), Can. J. Microbiol. **25**:1045–1051. Tapwater at pH 3.5 with 0.0005 M aluminum chloride was seeded with poliovirus and passed through a 0.45- + 0.25-μm Filterite filter in a 47 mm holder. Virus was eluted by passing 20 ml of 3% pH 9 lysine through the filters. This eluate was then passed through a 25 mm Zeta plus C-10 filter. Virus adsorbed to the Zeta plus filter was recovered by passing 4 ml of 3% pH 9 beef extract through the filter.

TABLE 5.9. Poliovirus Type 1
Concentration from Tapwater by Capillary
Ultrafiltration[a]

Experiment Number	Total Virus Input PFU	Virus Recovery (%)
1	150	79
2	200	115
3	45	66
4	39	74
5	41	67
	Mean:	80

[a] Adapted from Y. Rotem-Borensztajn, E. Katze-nelson, and G. Belfort (1979), *J. Environ. Eng. Sci.* **105**:401–407.

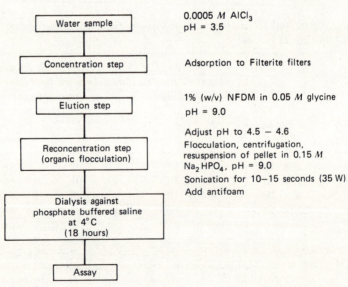

FIG. 5.11. Scheme for recovering viruses by the non fat dry milk (NFDM) technique. From G. Bitton et al. (1979), *Water, Air, Soil Poll.* **12:**187–195.

Nonfat dry milk is an inexpensive proteinaceous eluent that performs as well as casein in virus recovery from membrane filters. A nonfat dry milk technique has been proposed recently for virus recovery from seawater or tapwater. Details on this method are given in Figure 5.11.

Virus Concentration from Sewage

Raw sewage and wastewater treated effluents harbor much higher numbers of viruses than either tapwater or seawater. Therefore relatively small volumes of raw sewage (1–5 liters) are sufficient for virus detection. Membrane filtration technology allows the processing of up to 50 gallons (189 liters) of wastewater treated effluents. One particular feature of sewage is the presence of solid-associated viruses (Table 5.3). These viruses may be embedded within flocs or adsorbed on the surface of sewage solids. Available technology (e.g., Wallis-Melnick and EPA virus concentrators) allows an overall recovery of 50% of imput virus, including the surface-adsorbed viruses. However, the recovery of embedded viruses remains to be demonstrated.

Virus Detection in Surface Waters

Membrane filtration technology has also allowed the detection of viruses in surface waters. A survey of four rivers in France (Marne, Moselle, Rhin, Seine) has revealed virus concentrations ranging from 40 to 120 PFU/l (Table 5.10)

A Unified Scheme for Virus Concentration from Environmental Waters

A unified scheme has been developed for virus (poliovirus type 1) concentration from tapwater, seawater, and sewage. Details concerning the methodology

TABLE 5.10.   Virus Detection in Surface Waters[a,b]

River	Number of Samples	Virus (PFU/liter)
Marne	2 samples of 20 liters	40
Moselle	6 samples of 20 liters	40
Rhin	6 samples of 20 liters	56
Seine	3 samples of 200 liters	120

[a] Adapted from J. C. Block et al. (1978), *TSM Eau*, March 1978, p. 181–184.

[b] Concentration method: Adsorption onto fiberglass filters in the presence of 5×10^{-4} M $AlCl_3$ at pH 3.5. Elution with glycine—NaOH at pH 11.5.

TABLE 5.11. Unified Scheme for the Recovery of Poliovirus Type I from Environmental Waters[a]

| Type of Water | I. Concentration | | | II. Elution | | III. Reconcentration | | Average Virus Recovery % |
	AlCl₃ Conc. (M)	pH	Adsorbent Filters	pH	Times Eluent Passed Through Filters	1–100-Gal Sample	100–500 Gal Sample	
Tapwater	0.0005	3.5	Fiberglass depth filter (K27) and/or 0.25-μm porosity pleated filter	10.5 or 11.5	5 or 1	Membrane filters	AlCl₃ flocculation followed by hydroextraction	41–82 (Mean = 52)
Seawater	0.0015	3.5	Fiberglass depth filter (K27) or 3.0-μm porosity pleated filter and 0.45-μm porosity pleated filter	11.5	1	FeCl₃ flocculation	AlCl₃ flocculation followed by hydroextraction	42–63 (Mean = 53)
Sewage	0.0005	3.5	Fiberglass depth filter (K27) or 3.0-μm porosity pleated filter and 0.45-μm porosity pleated filter	11.5	1	AlCl₃ flocculation followed by hydroextraction		32–70 (Mean = 50)

[a] Adapted from C. P. Gerba et al. (1978), *Appl. Environ. Microbiol.* **35:**540–548.

employed are given in Table 5.11. It is possible to recover approximately 50% of input virus from any of the tested waters.

5.4 DECONTAMINATION OF VIRUS CONCENTRATES

Once a water or wastewater sample has been concentrated, it is necessary to decontaminate the concentrate prior to viral assay. Decontamination is the inactivation or removal of bacteria and fungi that may destroy the tissue culture during virus assay. The contamination problem is especially acute during the processing of sewage effluents or heavily polluted waters.

Various methods of decontamination have been considered and the most important ones are:

1. *Antibiotics.* Antibiotics, at concentrations higher than those generally recommended for tissue culture, are widely employed to tackle the contamination problem. The most successfully used antibiotics are penicillin, streptomycin, fungizone, and mycostatin. Once treated with antibiotics, the virus sample must be incubated for some hours (3 hours) at 37°C before assay.

2. *Membrane Filtration.* This method is successful only after treating the membrane filter with a substance (fetal calf serum, beef extract, Tween 80) which prevents the adsorption of the viruses to the filter. Moreover, it is not recommended if solids are present.

3. *Ether* or *Chloroform.* These solvents have the ability to kill bacteria and fungi without harming enteric viruses (viruses with a lipid envelope are inactivated). Ether, due to risks associated with its use, is not as widely accepted as chloroform. The latter is also conveniently separated below the water phase. After proper contact time (0.5 to 1 hour at room temperature or refrigeration overnight) the sample may be centrifuged and the top aqueous phase removed and subsequently aerated for some hours to remove the remaining traces of chloroform.

5.5 VIRUS ASSAY

5.5.1 General Remarks

Presently there is no available universal method used for the detection of the more than 100 enteric viruses known to exist. Most of these viruses can be detected by using host cell lines of human (e.g., HeLa or amnion cells) or simian (e.g., Buffalo green monkey (BGM) or vero cells) origin. Some others, such as coxsackieviruses A, are detected by inoculation of newborn mice. For the virus responsible for infectious hepatitis A there is no available tissue culture assay method.

The growth rate of virus is also an important factor to be considered in the detection methodology. For example, adenoviruses grow very slowly and are thus detected only after 3–4 weeks.

5.5.2 Importance of the Type of Host Cell in the Recovery of Enteric Viruses

Since tissue culture is the most important tool used in virus assay, it is necessary to find a cell line which is sensitive to most enteric viruses present in environmental samples. BGM, a continuous cell line established from primary

TABLE 5.12. Influence of the Type of Host Cell Line on the Recovery of Enteric Viruses from Natural Water and Wastewater

	PFU Recovered		
Type of Water	BGM[a]	Vero	Primary Rhesus Kidney
Wastewater from Haifa, Israel[b]	66,000	5600	—
Trickling filter effluent[b] from Haifa, Israel	8,500	4000	—
Raw sewage (USA)[c]	93	—	35
	22	—	20
	1,285	—	73
Chlorinated primary effluent[c]	660	—	279
	1,082	—	41
	730	—	28
Ocean water[c]	4	—	1
	3	—	0
River water[c]	18	—	0
Storm sewer at river	2,703	—	32

[a] BGM: Continuous cell line established from primary African green monkey (*Cercopithecus aethiops*) kidney cells.

[b] Data from N. Buras (1976), *Water Res.* **10**:295. The numbers represent the PFU recovered per 100 ml.

[c] Data from D. R. Dahling et al. (1974), *News of Environ. Res. in Cincinnati*. The numbers represent the PFU recovered per 2 liters.

African green monkey (*Cercopithecus aethiops*) kidney cells, is generally more sensitive to a variety of enteric viruses than other simian cells (vero or primary Rhesus kidney cells). Table 5.12 shows that in wastewater and surface water the number of viruses recovered is always higher with the BGM cells. This continuous cell line is unfortunately less sensitive for the isolation of certain echoviruses (e.g., echo 11 and 18) and adenoviruses (e.g., adenovirus 1, 2, and 11), and loses its sensitivity for viruses after a certain number of passages.

5.6   ENUMERATION AND DIAGNOSTIC PROCEDURES FOR VIRUSES IN ENVIRONMENTAL SAMPLES

Once viruses have been adequately concentrated and once a suitable host cell line has been selected, one proceeds to their enumeration and identification. This topic has been discussed in Chapter 1.

5.7   SUMMARY AND CONCLUDING REMARKS

The choice of a concentration and detection method is dictated by many factors:

1. *Water Quality.* Environmental waters vary widely in their physico-chemical characteristics such as pH. organic material content, salt concentration, suspended solids, color content, etc. Therefore, one some-times needs to adjust some of these parameters in order to concentrate viruses. The solid-associated viruses have to be taken into account in any concentration scheme. One generally use prefiltration to avoid the problem of membrane clogging, and it is recommended that the prefilter should also be processed for virus recovery.

2. *Volume of Water.* The choice of a concentration method should also be dictated by the amount of water to be processed. The Wallis-Melnick and EPA concentrators, based on adsorption to and elution from membranes, are indicated for large volumes of waters. Procedures such that two-phase separation, hydroextraction or adsorption to precipitable salts, are rather indicated for small volumes of grossly polluted waters.

3. *Elution of Viruses.* We have seen that most of the concentration methods are based on the concept of adsorption to surfaces. In these particular techniques the recovery of viruses from the surface with an appropriate eluent (generally a proteinaceous material at a high pH) is as important as the adsorption step. The development of methods that involve moderately alkaline eluents (e.g., pH 9) should be encouraged.

4. *Virus Assay.* There is no one host system which is able to recover all the enteric viruses present in water.

5.   *Time.* The concentration and detection of viruses may take many days to be completed. The concentration procedure itself generally takes 1 to 24 hours. We, therefore, need more rapid methods for the detection of viruses in environmental samples.

Environmental virologists are still actively searching for the ideal method which would be rapid, easy, applicable to all types of waters, sensitive, reproducible, and inexpensive. This technological challenge has to be resolved before imposing virus standards for water quality.

5.8   FURTHER READING

Bitton, G. 1975. Adsorption of viruses onto surfaces in soil and water. *Water Res.* **9**:473–484.

Foliguet, J. M., J. Lavillaureix, and L. Schwartzbrod, 1973. Virus et eau: II: Mise en évidence des virus dans le milieu hydrique. *Rev. Epidemiol. Méd. Soc. Santé Publ.* **21**:185–259.

Goyal, S. M., and C. P. Gerba. 1979. Membrane filters in virology. In: B. J. Dutka, Ed., *Membrane Filtration: Techniques, Applications and Problems.* Marcel Dekker, New York.

Hill, W. F., Jr., E. W. Akin, W. H. Benton. 1971. Detection of viruses in water: A review of methods and application. *Water Res.* **5**:967–995.

Katzenelson, E. 1974. Virologic and engineering problems in monitoring viruses in water. In: *Viruses in Water.* G. Berg et al., Eds. American Public Health Association, Washington, D.C.

Lund, E. 1977. Sampling and isolation methods for the detection of viruses in municipal wastewater. In: *Wastewater Renovation and Re-use,* F. M. D'Itri, Ed. Marcel Dekker, New York.

Marshall, K. C. 1976. *Interfaces in Microbiol Ecology.* Harvard University Press, Cambridge, Mass.

Sobsey, M. D. 1974. Methods for detecting enteric viruses in water and wastewater. In: *Viruses in Water,* G. Berg et al., Eds. American Public Health Association, Washington, D.C.

six
Fate of Viruses in Sewage Treatment Plants

6.1   INTRODUCTION: OCCURRENCE OF PATHOGENS IN WASTEWATER

We have seen in Chapter 1 that domestic wastewater may contain more than 100 types of enteric viruses. Viral numbers vary considerably with the season of the year and with the geographic location. It has been estimated that raw sewage may contain from 50 to almost 10^5 PFU/l. Most of the viral concentrations reported in the literature are based on the assumption that 100% of the observed plaques are of viral origin. Recent studies suggest, however, that less

TABLE 6.1.   Major Bacterial Pathogens, and Protozoan and Helminth Parasites Present in Domestic Wastewater

Organism	Disease Caused
Bacteria	
Vibrio cholera	Acute intestinal disease
Salmonella	*S. typhi*: Agent of typhoid fever
Shigella	Agent Bacillary dysentery
Mycobacterium	Tuberculosis
Leptospira	Causes infections involving liver, kidney, and central nervous system
Enteropathogenic *E. coli*	Gastroenteritis
Protozoa	
Entamoeba histolytica (amoeba)	Ulcers
Giardia lamblia (flagellate)	Diarrhea, anorexia, abdominal pains
Balantidium coli (ciliate)	Ulcers, dysentery
Naegleria (amoeba)	*N. fowleri*: meningoencephalitis
Helminthes	
1. *Nematodes* (roundworms)	
Ascaris lumbricoides	Intestinal obstruction in children
Necator americanus	Anemia
Ancylostoma duodenale	Anemia
2. *Cestodes* (tapeworms)	
Taenia saginata	Abdominal discomfort, hunger pains, chronic indigestion
3. *Trematodes* (Flukes)	
Schistosoma mansoni	Complications in liver (cirrhosis), bladder and large intestine (Not a problem in the United States because of absence of host snail)

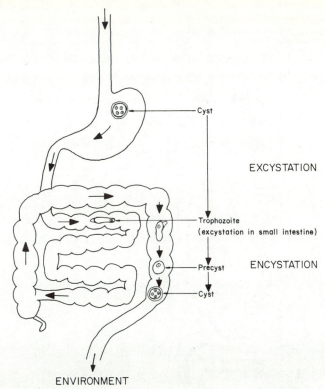

FIG. 6.1. Life cycle of *Entamoeba histolytica*. Adapted from H. C. Jeffrey and R. M. Leach (1972), *Atlas of Medical Helminthology and Protozoology*, Churchill Livingstone, Edinburg.

than 20% of the plaques could be confirmed as of viral origin and that the virus concentration of sewage samples was frequently overestimated. An examination of sewage samples from seven wastewater treatment plants in Illinois, Tennessee and Ohio has revealed that sewage contained from 0 to 80 confirmed plaques/liter.

Domestic wastewaters harbor a wide range of bacterial pathogens as well as protozoan and helminth parasites. The microbe and parasite content of domestic sewage generally reflects the range of enteric infections encountered in the community. The major groups of sewage pathogens are listed in Table 6.1.

Bacterial pathogens in wastewater include *Salmonella, Shigella*, enteropathogenic *E. coli, Mycobacterium*, and *Leptospira. Salmonella* is the most common pathogen found in wastewater. *S. typhi* is the agent responsible for typhoid fever, a serious disease. *Shigella* is responsible for outbreaks involving

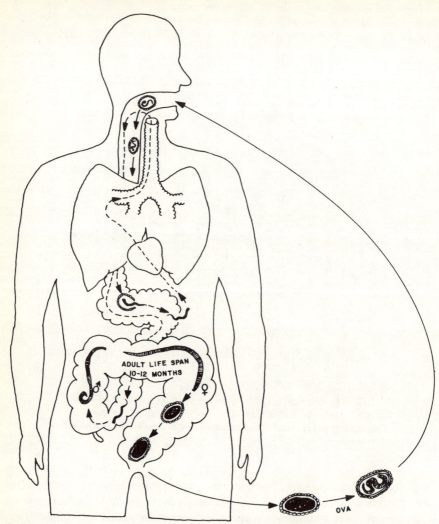

FIG. 6.2.   Life cycle of *Ascaris lumbricoides* (roundworm). Adapted from H. C. Jeffrey and R. M. Leach (1972), *Atlas of Medical Helminthology and Protozoology*, Churchill Livingstone, Edinburgh.

thousands of cases. Enteropathogenic *E. coli* causes outbreaks of gastroenteritis.

Protozoan parasites include members of the amoeba (*Entamoeba histolytica, Naegleria fowleri*), ciliate (*Balantidium coli*), and flagellate (*Giardia lamblia*) groups. Upon their release into the environment they form *cysts,* which may be transmitted via wastewater, flies, infected food handlers, or direct contact. Upon entry into the human host, the cysts undergo *excystation* and liberate active forms called *trophozoites.* It is believed that 10 to 20 cysts constitute an infective dose. The life cycle of a typical protozoan parasite, *Entamoeba histolytica,* is shown in Figure 6.1. This protozoa causes ulcers and diarrhea that may result in the death of the host. Epidemiological surveys in Colorado have shown that *Giardia lamblia* affects mountain hikers that consume untreated stream water.

It is estimated that at least half of the world population is infected by helminth (worms) parasites. They invade the alimentary canal as well as other parts of the body (liver, muscles, nervous system, skin, etc.). They cause major complications ranging from anemia and blindness to intestinal obstruction.

The helminthes of medical importance are subdivided into three classes:

- Cestoda: Tapeworms.
- Trematoda: Flukes.
- Nematoda: Roundworms.

Examples for each class are given in Table 6.1 along with the diseases caused. The infective stage of parasitic helminthes are the *ova* (eggs), which are excreted in feces and spread via wastewater, soil, or food. The ova are very resistant to environmental stresses and to chlorination in sewage treatment plants. The life cycle of a typical helminthe, *Ascaris lumbricoides* (nematode), is illustrated in Figure 6.2.

6.2 MAJOR STEPS INVOLVED IN A SEWAGE TREATMENT PLANT

We have discussed the presence of viruses, bacterial pathogens, and protozoan and helminthic parasites in domestic wastewater. The function of a sewage treatment plant is to remove or at least reduce these undesirable biological components along with suspended solids, organic materials (dissolved and particulate matter), recalcitrant organic compounds such as pesticides, inorganic nutrients such as nitrogen and phosphorus, and heavy metals.

A complete sewage treatment plant comprises the following major steps, which are illustrated in Figure 6.3:

1. *Primary treatment.* It is a physical process which involves the separation of coarse debris, followed by sedimentation.

2. *Secondary treatment.* It is essentially a biological process which is carried out by a mixture of bacteria, fungi, algae, protozoa, rotifers, worms, and insect larvae. There are three major approaches to biological treatment: activated sludge, trickling filters, and oxidation ponds.

   A disinfection step is generally included at the end of a biological treatment process. This important step will be discussed in Chapter 8.

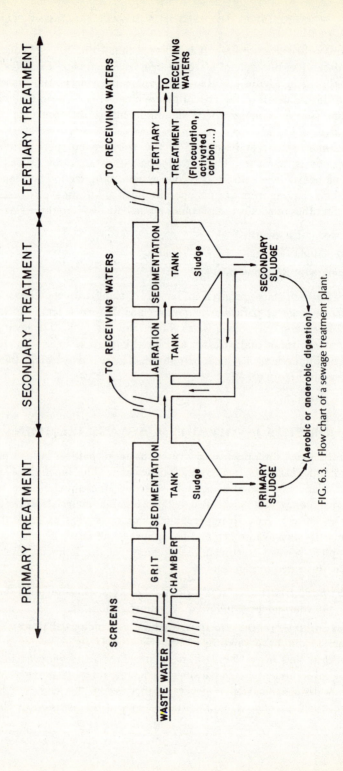

FIG. 6.3.   Flow chart of a sewage treatment plant.

126

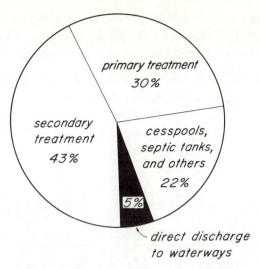

FIG. 6.4.   Sewage treatment in the United States in 1972 by percent of the population. Adapted from G. T. Miller, Jr. (1975), *Living in the Environment: Concepts, Problems and Alternatives,* Wadsworth, Belmont, Calif.

Similarly, the subject of aerobic and anaerobic digestion of primary or secondary sludge will be examined separately in Chapter 10.

3. ***Tertiary treatment.*** The prime function of this physicochemical treatment is to further remove some nutrients (N, P), some dissolved organic matter, heavy metals, and bacterial and viral pathogens.

Figure 6.4 shows sewage treatment in the United States by percent of the population. It is evident that the trend is leaning toward secondary waste treatment (43% of the population is served by secondary waste treatment). Public law 92-500 requires the establishment of biological treatment by the end of 1977 in municipal wastewater treatment plants.

We will now describe in more detail the design and biology of the various steps in a sewage treatment plant and discuss their virus-removal capacity.

6.3   FATE OF VIRUSES IN SEWAGE TREATMENT PLANTS

6.3.1   Association of Viruses with Sewage Solids

Domestic wastewater contains a variety of solids of biological and mineral origin. New solids are produced during the transformation of soluble organics to microbial cells. A virus may be adsorbed on the surface of these solids or embedded within flocs. Therefore the proper detection of viruses in sewage should include some drastic treatment, such as sonication, in order to free some of the embedded virus particles. Solids-associated viruses may be demonstrated by passing wastewater or wastewater effluents through calf serum-treated membrane filters that allow only the passage of single virus particles.

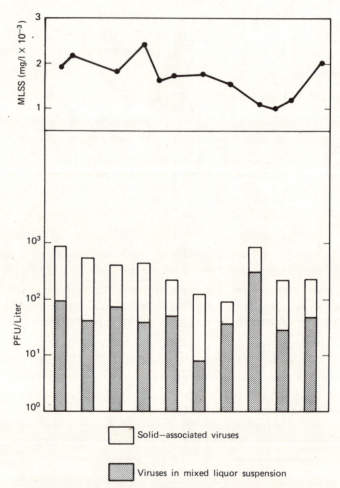

FIG. 6.5.   Virus Association with Wastewater solids. (Dark bars are enteric virus levels in the liquid phase of mixed liquor while open bars represent solid associated viruses.) From B. E. Moore et al. (1974), In: *Virus and Survival in Water and Wastewater Systems*. J. F. Malina, Jr. and B. P. Sagik, Eds., University of Texas at Austin, Texas. Courtesy of the Center for Research in Water Resources, The University of Texas at Austin.

TABLE 6.2.   Removal of Enteric Viruses by Primary and Activated Sludge Treatments at Dadar Sewage Treatment Plant, Bombay, India[a]

Season	Months	Raw Sewage: Virus Concentration PFU/Liter	% Reduction After Primary Treatment	% Reduction After Activated Sludge Treatment
Rainy				
	June 1972	1000	33.5	97.9
	July 1972	1250	24.1	97.0
	June 1973	1200	29.7	95.5
	July 1973	837	29.8	98.9
Autumn				
	September 1972	300	64.7	98.0
	October 1972	312	66.0	91.7
	October 1973	572	73.0	96.4
	November 1973	1087	56.0	90.0
Winter				
	January 1973	587	41.4	96.8
	February 1973	468	83.4	95.5
	January 1974	605	47.0	99.6
Summer				
	March 1973	812	57.0	97.6
	April 1973	875	59.7	98.6
	May 1973	731	66.0	93.5
	March 1974	250	68.8	98.0
	June 1974	694	74.7	99.0

[a] Adapted from Rao et al. (1977), *Prog. Water Technol.* **9**:113–127.

   It was found that 3 to 90% of viruses in wastewater may be solids-associated. Figure 6.5 illustrates the proportion of "free" and solids-associated viruses in a wastewater with a concentration of solids ranging from 980 to 2400 mg/l. The levels of solid-bound viruses were 5- to 10-fold higher than those of "free" viruses. During waste treatment, some of these bound viruses may be released following disruption of flocs or changes in physicochemical conditions.

   Virus monitoring in sewage must take into consideration these solid-associated viruses. Failing to do so might lead to erroneous results and to a false sense of security. Moreover, one must add that the solid-bound viruses are as infective as "free" viruses. This subject is examined in Chapter 4.

6.3.2   Primary Treatment

Primary treatment is a physical process that includes:

1.   *Screening.* It removes sticks and other materials that may plug pumps and pipes.
2.   *Grit removal.* It removes heavy materials such as gravel or bone chips.
3.   *Primary sedimentation.* It is accomplished in a tank where the "settlable" solids and floating materials are eliminated. The detention time in the tank is between 1.5 and 3 hours. This process produces a *primary sludge* that contains 2 to 6% solids, and which is further digested anaerobically.

     Primary sedimentation removes bacterial pathogens ineffectively. The removal is approximately between 20 and 40%. Similarly, protozoan cysts are not removed effectively by this process. However, eggs of *Ascaris* and *Taenia saginata* settle fairly well within a 3-hour sedimentation period.

It has been shown on many occasions that the primary treatment of domestic wastewaters plays a minor role in virus removal. The reduction of viruses may range between 0 and 30% during a 3-hour settling period. Unfortunately, many investigators, using seeded viruses, failed to take into account the association between viruses and sewage solids. It may be that this association will result in a higher removal of viruses, particularly those embedded within sewage solids. Field studies in India have shown that enteric virus removal during primary treatment ranged between 24 and 33% during the monsoon (rainy season) and between 41 and 83% during the other seasons (Table 6.2).

A settling time beyond the usual 2 to 3 hours would increase the efficiency of primary treatment for virus removal.

6.3.3   Activated Sludge

Description of the Process

An activated sludge system is essentially an aerobic biological process which transforms soluble organic materials to microbial biomass. This process was first used in England at the end of the nineteenth century and was introduced in the United States during World War I.

A conventional activated sludge process includes (Figure 6.6) the following:

1.   An *aeration tank.* Primary effluent is introduced and mixed with returned sludge to form the *mixed liquor,* which contains 1500 to 2500 mg/l of suspended solids. (MLSS = mixed liquor suspended solids). Aeration is provided by mechanical means. The detention time in the aeration basin varies between 4 and 8 hours.

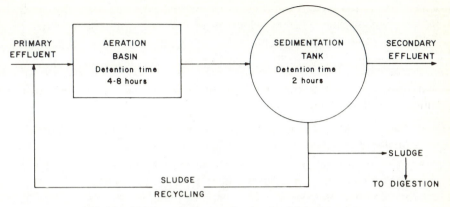

FIG. 6.6.   Diagram of a conventional activated sludge system.

2. A *sedimentation tank* (thickening tank). This is used for the settling of
    microbial flocs (sludge) produced during the aeration phase. A portion of
    the generated sludge is recycled back to the aeration basin. This sludge,
    which contains a wide variety of active microorganisms, has been called
    "activated sludge." The remaining sludge (waste activated sludge) is
    treated by aerobic or anaerobic digestion, and this topic is covered in
    Chapter 10. A properly designed activated sludge process usually results
    in a 90 to 95% removal of biochemical oxygen demand (BOD). There
    are, however, variations of the standard activated sludge process:
    (a) *Contact stabilization.* It is similar to the standard process
         except that the sludge is reaerated prior to mixing with the
         primary effluent.
    (b) *Extended aeration.* The aeration time is extended up to 24
         hours. This version is used in small installations.

Biology of Activated Sludge

A low BOD effluent is the result of two main processes occurring in activated sludge:

1. Oxidation of organic matter in the aeration basin.
2. Settling of microbial flocs in the sedimentation tank.

Activated sludge harbors a wide variety of microorganisms that participate in
the stabilization and settling processes.

Bacteria: Over 300 strains of bacteria have been found in activated sludge.
They are responsible for the oxidation of organic matter and they produce
polysaccharides and other polymeric materials (polypeptides, nucleic acids)

which help in the flocculation process. One of the most important floc-forming bacteria is *Zooglea ramigera* which produces abundant extracellular slime.

This flocculation process plays a significant role in the removal of pathogenic microorganisms by the activated sludge process.

Protozoa. Protozoa constitute a significant component of the activated sludge microflora. They are divided into:

- *Sarcodina* (Amoeba). They move by means of *pseudopods* ("false feet") and feed on soluble or solid organic particulates.
- *Mastigophora* (Flagellates). They move by means of flagella and feed on soluble (holophytic mastigophora) or solid food (holozoic mastigophora).
- *Ciliata*. They move by means of cilia. There are *free-swimming* and *stalked* forms which attach to sludge particles.

It is widely recognized that a knowledge of species composition of activated sludge may give a clue on the BOD removal efficiency of the process. It may also help in the detection of some drastic changes in the plant operation. There is an ecological succession of the various microbial and metazoan components during the activated sludge process. This succession is illustrated in Figure 6.7. One can see, for example, that stalked ciliata and rotifers indicate a low BOD and that the plant operation is satisfactory.

Filamentous Microorganisms. These are sheathed bacteria (*Sphaerotilus natans, Thiothrix sp.*) gliding bacteria (*Beggiatoa, Vitreoscilla*), and fungi (*Geotrichum candidum*), which are responsible for *sludge bulking*, a condition characterized by a poor settling of the sludge in the sedimentation basin. This nuisance condition is controlled by proper chlorination or by addition of lime, coagulants, or hydrogen peroxide.

Fate of Viruses and Other Pathogens in the Activated Sludge Process

The knowledge of the design and biology of activated sludge will help the reader in a better understanding of the removal of pathogens by this process.

It appears that protozoan cysts (e.g., cysts of *Entamoeba histolytica*) and metazoan eggs (e.g., eggs of *Ascaris*) are not efficiently removed by the activated sludge process. However, for some eggs the removal may be as high as 99%, and this probably occurs during the settling phase rather than the aeration phase. The prevailing conditions in activated sludge may even promote the hatching of some helminth eggs. Bacterial pathogens are well removed by the activated sludge treatment, and reductions of 90 to 95% have been reported. Their disappearance is due to physical and chemical factors, grazing by

High BOD

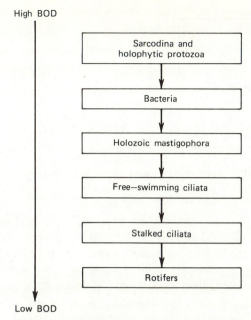

FIG. 6.7.   Ecological succession of activated sludge organisms. Adapted from R. E. McKinney and A. Gram (1956), *Sew. Ind. Wastes* **28:**1219.

protozoa (this has been shown by fluorescence microscopy) and to adsorption onto the surface and embedding within sludge flocs. This association results in their transfer to the sludge.

Activated sludge appears to be the best biological treatment process for virus removal. In laboratory experiments, bench model continuous flow-activated sludge units were able to remove 90 to 98% of poliovirus 1 and coxsackievirus A9. Plant scale studies have substantiated the data found in the laboratory. At the Dadar sewage treatment plant (Bombay, India), which processes sewage at a rate of 5 mgd (million gallons/day), 90 to 99% of enteric viruses were removed by a conventional activated sludge process (Table 6.2).

The removal of viruses by activated sludge may be due to the following:

1. *Adsorption of viruses to sludge solids.* Viruses become associated with sludge flocs made of microbial cells along with organic and inorganic debris. This association may be due to adsorption and/or encapsulation of viruses within the sludge flocs. Viral adsorption conforms to the Freundlich isotherm as shown in Figure 6.8. The removal by adsorption

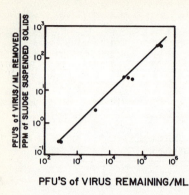

FIG. 6.8. Virus adsorption to sludge solids. Adapted from N. A. Clarke et al. (1961), *Am. J. Publ. Health* **51**:1118–1129.

is responsible for approximately 1 log reduction in virus titers. Thus, as shown for bacterial pathogens, the activated sludge process brings about a transfer of viruses from wastewater to sludges. The sludge-associated viruses are not completely inactivated and this should be considered when one contemplates some method of sludge disposal.

2. **Inactivation by sewage bacteria.** *Flavobacterium, Aerobacter,* and *Klebsiella* are three bacteria which appear to possess some antiviral activity during the activated sludge process.

3. **Ingestion by protozoa (ciliates) and small metazoa (nematodes).** (This topic is discussed in Chapter 3.)

4. **Inactivation by aeration.** This mode of inactivation was not found to be significant in viral reduction.

Whatever the cause of removal or inactivation one may conclude that activated sludge is relatively efficient both in the laboratory and under field conditions. However, virus may escape treatment following some problems in plant operation. We have already seen that bulking sludge results in poor effluents containing many flocs, which probably harbor many viruses. It thus appears that the capacity of an activated sludge to remove viruses is related to its ability to remove flocs and suspended solids.

6.3.4   Trickling Filters

A trickling filter system consists of a *filter* (Figure 6.9) which is a 4- to 8-feet deep tank packed with gravel, rocks, small stones, or plastic materials. The tank is circular and has a rotary distributor that discharges the primary effluent on the surface of the filter bed. The spraying or "trickling" action of the distributor helps saturate the sewage with oxygen.

There are two main types of trickling filters:

- *Low-rate trickling filters.* The organic loading rate is from 10 to 20 lb BOD/1000 ft³.
- *High-rate trickling filters.* The loading rate is around 90 lb BOD/1000 ft³. This category generally involves the recirculation of the settled effluent.

The trickling of sewage on the surface of the stones results in the formation of a microbial film called *zoogleal film.* This biological slime layer is composed of bacteria (which secrete extracellular polysaccharides), fungi, protozoa, and algae. The organic matter is biooxidized as it trickles over the surface of zoogleal film. The buildup of the slime layer creates an anaerobic zone and results in the sloughing of the old slime, which is collected in the final clarifier. Figure 6.10 describes the zoogleal film in a trickling filter. Since this layer is important in the removal of BOD, the control of its growth is of utmost importance for the proper functioning of a trickling filter. Diptera larvae belonging to the genus *Psychoda* help in that control by eating part of the slimy material and, thus, help avoid clogging of the bed. When they reach the adult stage these "filter flies" may be a nuisance to sewage treatment operators. Other organisms involved in the control of zoogleal film include rotifers and nematodes.

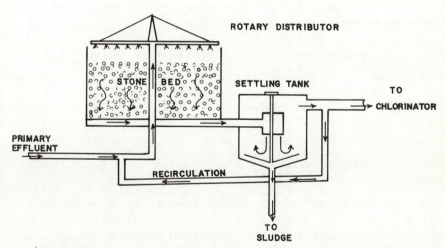

FIG. 6.9. The trickling filter process. Adapted from P. M. Higgins (1968), *Dev. Ind. Microbiol.* **9**:146–159.

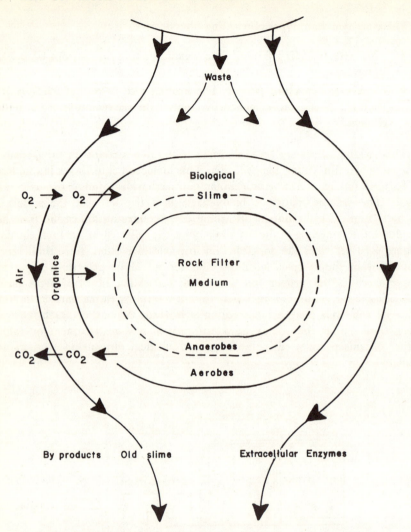

FIG. 6.10. Zoogleal film formation on packing media in trickling filters. From J. E. Zajic (1971), *Water Pollution: Disposal and Re-use.* Marcel Dekker, New York. Courtesy of Marcel Dekker, Inc.

The removal efficiency (BOD and suspended solids) of trickling filters is usually less than that obtained in the activated sludge system. However, these filters are easier to operate and are widely used by industry.

The trickling filter system does not completely remove bacterial pathogens, protozoan cysts, and helminth eggs. Moreover, the removal pattern is generally variable (Table 6.3). However, it was reported that 88 to 99% of cysts of *Entamoeba histolytica* are removed by this process. It has been found that conditions prevailing in activated sludge or trickling filters may promote the hatching of some eggs.

The removal of viruses by trickling filters is generally low and inconsistent. The first field experiments have shown that virus removal was around 40%. Unfortunately, a great deal of field monitoring was essentially qualitative. However, there has been some quantitative studies on the fate of RNA phage f2 in two conventional trickling filters. Figure 6.11 shows that the trickling filter (prior to settling in the secondary clarifier) removed less than 20% of added virus in both plants. The cumulative virus removal after clarification and chlorination was less than 90% for both plants.

In laboratory experiments, rotary tube-type trickling filters have been used to assess their efficiency in virus removal (poliovirus 1, coxsackievirus A9, and echovirus 12). Table 6.4 shows the performance of two trickling filters operating at medium (10 mgd/acre) and high (23 mgd/acre) filtration rates. It is clear that the medium filtration rate unit was more efficient in virus removal as well as in the reduction of coliforms, fecal streptococci, BOD, and COD. Poliovirus 1, echovirus 12, and coxsackievirus A9 were reduced by 85%, 83%, and 94%, respectively, in trickling filters operating at 10 mgd/acre.

It is believed that virus removal by trickling filters is an adsorption phenomenon and that the adsorbed virus is inactivated when coming into contact with the slime layer.

TABLE 6.3. Removal of
Pathogens by Trickling Filters[a]

Pathogen	% Removal
Coliforms	82–97
Salmonella	84–99.9
Mycobacterium	66–99
Amoebic cyst	11–99.9
Helminth ova	62–76

[a] Adapted from D. H. Foster and R. S. Engelbrecht (1970), in: *Recycling Treated Municipal Wastewater and Sludge Through Forest and Cropland*, W. E. Sopper and L. T. Kardos, Eds. Pennsylvania State University Press, University Park, Pa.

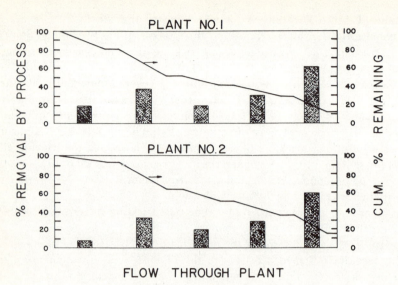

FIG. 6.11. Removal of phage f2 by trickling filters. Adapted from V. R. Sherman et al. (1975), *Water Sew. Works* **122:**R36.

TABLE 6.4. Removal of Enteric Viruses by Rotary-tube Trickling Filters in Laboratory Experiments[a]

	Virus	Virus Reduction %	Coliform Reduction %	Fecal Strepto-cocci Reduction %	BOD Reduction %	COD Reduction %
Medium	Poliovirus 1	85	94	88	92	94
Filtration	Echovirus	83	95	96	94	95
Rate	12					
(10 MGD/	Coxsackie-	94	94	91	93	95
acre)	virus A9					
High	Poliovirus 1	59	75	75	72	57
Filtration	Echovirus	73	44	62	75	58
Rate	12					
(23 MGD/	Coxsackie-	81	66	75	70	60
acre)	virus A9					

[a] Adapted from N. A. Clarke and S. L. Chang (1975), *Appl. Microbiol.* **30:**223–228.

6.3.5   Oxidation Ponds

Oxidation or stabilization ponds are one of the oldest methods of biological treatment of wastewater and are used in more than 40 countries. Due to land requirements they are commonly used in rural areas. There are many categories of ponds, namely, aerobic, anaerobic, and aerated lagoons. In the latter, aeration is provided by mechanical means. An oxidation pond is usually 3–5 feet deep, and this depth allows maximum algal growth. The detention time ranges between 1 and 4 weeks and is an essential parameter in waste degradation.

The treatment of wastes in oxidation ponds is the result of natural biological processes carried out mainly by bacteria and algae. A diagram of major biological events occurring in an oxidation pond is shown in Figure 6.12. Oxygen produced as a result of algal photosynthesis is used by heterotrophs to degrade the organic wastes entering the ponds. Carbon dioxide is in turn produced as a result of bacterial metabolism and is used for algal photosynthesis. Hence there exists a beneficial relationship between bacteria and algae. Dead algal, bacterial, and fungal cells settle to the bottom of the pond. The settled "sludge" is

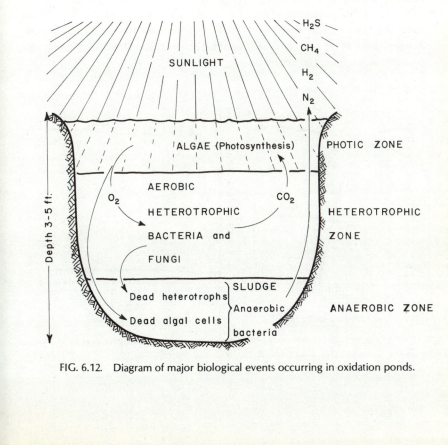

FIG. 6.12.   Diagram of major biological events occurring in oxidation ponds.

in turn decomposed by anaerobic bacteria, resulting in the production of gases such as methane, hydrogen sulfide, and nitrogen gas.

A significant portion of indicator and pathogenic bacteria is removed following passage through an oxidation pond. Removals above 90% or even 99% have been reported for *E. coli* or *Salmonella* sp. The reduction of indicator bacteria (total coliforms, fecal coliforms, and fecal streptococci) may vary from 70 to more than 99%.

When addressing the phenomenon of bacterial reduction in oxidation ponds one must take into account the interaction(s) between the two main biological components in oxidation ponds: bacteria and algae. These interactions comprise the following:

- *Antibiosis.* Algal cells may release antibiotic substances against indicator or pathogenic bacteria.

- *Stimulation.* Algal excretions may enhance the growth of some indicator bacteria.

- *Antagonism.* It has been shown that the reduction of *E. coli* was due to the high pH (10–10.5) generated as a result of algal photosynthesis. However, the two biological components (*E. coli* and *Chlorella*) can grow together when the wastewater is buffered to pH = 7.5. This phenomenon is illustrated in Figure 6.13. In addition to the biological interactions there are physical factors (temperature, sunlight) which also contribute to bacterial reduction.

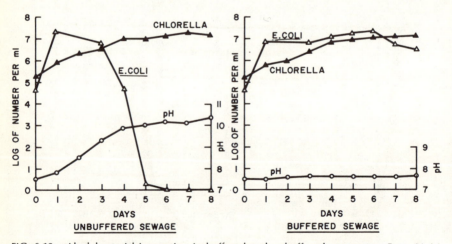

FIG. 6.13. Algal–bacterial interaction in buffered and unbuffered wastewater. From N. M. Parhad and N. V. Rao (1974), *J. Water Poll. Control Fed.* **46:**980. Courtesy of Water Pollution Control Federation.

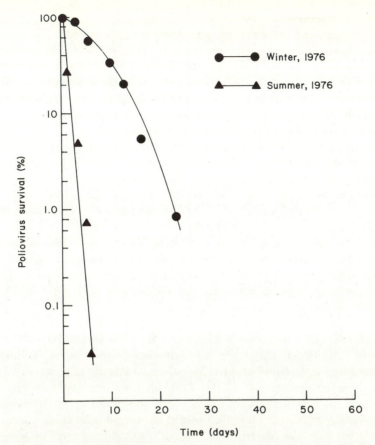

FIG. 6.14. Effect of the season of the year on the survival of poliovirus type 1 in a model oxidation pond. Adapted from L. Funderburg et al. (1978), *Prog. Water Technol.* **10:**619–629.

The ova (eggs) of helminth parasites are probably removed by settling to the sediments. Some ova may hatch under aerobic conditions prevailing in some ponds.

As in the case of bacterial pathogens, viruses are not completely removed by oxidation ponds. Field studies have shown erratic results and low virus removal in general, and the poor performance may have been due to short circuiting in the pond. However, laboratory scale oxidation ponds usually display higher removals under a wide range of BOD loading (100–550 lb BOD acre^{-1} day^{-1}).

We shall now analyze the various factors which may contribute to the inactivation and/or removal of viruses in oxidation ponds.

Physical Factors

Temperature and solar radiation are the most important factors controlling virus survival in this particular environment. A good virus removal would be expected under a hot and sunny climate in the upper layer of a shallow pond. Figure 6.14 displays the influence of the season of the year on the survival of poliovirus type 1 (Chat) in a model oxidation pond (150-cm diameter holding pond made of cast concrete). Virus numbers were reduced by 99% in 5 days during summer and 25 days during the winter season.

The adsorption of viruses to suspended solids and subsequent settling to the bottom of the pond can account for some reduction in virus concentration. Virus deposition in pond sediments is clearly illustrated in Figure 6.15. In this particular model pond, most of the viruses settled within a week period. It is recognized that in operational ponds viruses may settle even faster since most of the solids appear to reach the sediments within hours.

Virus associated with sediments survive for longer periods than in the water column. Any natural or artificial disturbance (spring mixing, irrigation pipes) of these sediments may result in virus increase in the pond effluent. This, of course, presents a great potential hazard, particularly when pond effluents are used in spray irrigation schemes.

Interaction of Viruses with Algae and Bacteria in Oxidation Ponds

Heavy growth of algal cells in oxidation ponds results in increased concentrations of dissolved oxygen. Moreover, as a result of photosynthesis the pH of the water can rise to 9.5–10.0. Probably dissolved oxygen does not affect virus survival. However, higher pH has been shown to increase virus inactivation in oxidation ponds. This effect would not probably be as significant as in the case of indicator bacteria (see Figure 6.13). The subject of algal–viral interaction remains to be thoroughly investigated.

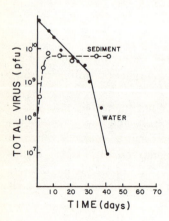

FIG. 6.15.   Poliovirus survival in oxidation pond sediments (winter season). From L. Funderburg et al. (1978), *Prog. Water Technol.* **10**:619–629. Courtesy of Pergamon Press, Inc.

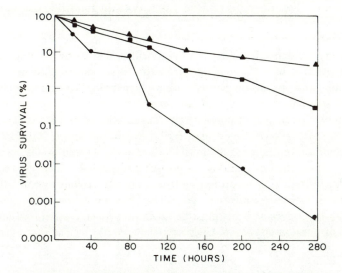

● — ● VIRUS AND POND BACTERIA

▲ — ▲ VIRUS AND ALGAE (S.QUADRICAUDA)

■ — ■ CONTROL (VIRUS SUSPENDED IN SEWAGE)

FIG. 6.16.   Effect of pond bacteria on virus inactivation in oxidation ponds. Adapted from M. D. Sobsey and R. C. Cooper (1973), *Water Res.* 7:669.

Bacteria have also been shown to increase viral inactivation in oxidation ponds. The addition of pond bacteria to a virus suspension results in more than $2 \log_{10}$ reduction in poliovirus numbers (Figure 6.16).

It is also probable that protozoa and small metazoa (rotifers, nematodes) ingest some viral particles but this mode of inactivation may not be significant. It is even possible that nematodes protect ingested viruses from the action of disinfectants. This phenomenon is of prime importance since nematodes are sometimes found in finished waters.

6.3.6   Tertiary Treatment of Domestic Wastewater

Virus Removal Capacity of Advanced Wastewater Treatment

Tertiary treatment comprises a series of physicochemical steps designed to further reduce pathogenic microbes, heavy metals, nitrate, phosphorus, trace organics, color, suspended solids, and turbidity. Most steps involved in advanced wastewater treatment have been used for many years in water treatment plants, and the reader is referred to Chapter 7 for their detailed description.

Various treatments are employed according to the intended use of the water. Table 6.5 illustrates the virus removal capacity of the following major steps.

Coagulation–sedimentation with Alum, Lime, or Polyelectrolytes. This process is intended to remove phosphorus and suspended solids. It is also the most effective way of removing viruses from secondary effluents. Alum or lime, added in appropriate concentrations, can bring about more than 99% reduction in virus numbers. An aluminum:phosphorus radio (Al:P) of 7 results in 99.7% removal of poliovirus 1. Coagulation with lime is also effective (99–99.9% removal) when its concentration is equal or above 500 ppm. At this level, the addition of lime results in pH values above 11 and this is known to be very detrimental to virus survival. Viruses are probably heavily reduced in lime-sludge, although some protection from high pH probably occurs due to the embedding of viruses within the sludge flocs. However, viruses are not inactivated in alum sludges and this must be taken into consideration in sludge treatment and disposal practices (see Chapter 10).

TABLE 6.5.   Virus Removal in Advanced Wastewater Treatment[a]

Treatment	Virus	% Removal
Coagulation–flocculation		
Alum	Polio 1	99.7
	Phage MS2	90–99.8
	Phage T4	96–97
Lime	Polio 1	99–99.9
Polyelectrolytes (cationic)	Phage T4	99.9
	Phage MS2	99–99.9
Sand filtration	Phage T4	99
	Polio 1	>99
Adsorption to activated carbon	Phage T2	75
	Phage T4	35–99
	Polio 1	90

[a] Adapted from J. F. Malina, Jr. (1976), in: *Virus Aspects of Applying Municipal Wastes to Land,* L. B. Baldwin et al., Eds. University of Florida, Gainesville, Fla.

TABLE 6.6.   Removal of Poliovirus Type 1 by
Lime Flocculation Followed by Sand Filtration[a,b]

Lime Concentration ppm Ca(OH)$_2$	pH of Final Effluent	Sampling Time[c] (min)	Virus Removal (%)
200	9.24	0	98.6
	9.27	12	98.8
	9.27	24	99.0
500	11.01	0	99.94
	11.10	12	>99.992
	11.10	24	99.98

[a] Adapted from G. Berg et al. (1968). *J. Am. Water Works Assoc.* **60**:193.

[b] The initial virus concentration was 33,333 PFU/l.

[c] Timed from beginning of filtration through sand.

Sand Filtration. A typical sand filter is described in Chapter 7. When properly used, sand filtration can achieve a good virus removal. This is best accomplished by combining sand filtration with some flocculation step. Table 6.6 shows the removal of poliovirus 1 by lime flocculation followed by sand filtration. In the presence of 500 ppm of lime, this particular process can result in 4 $\log_{10}$ reduction in virus concentration.

Activated Carbon Adsorption. Virus removal by activated carbon columns is generally low and variable. The performance of the columns depends on the flow rate and on the presence of soluble organics which compete with the viruses for adsorption sites on the activated carbon. However, activated carbon is nonetheless efficient in removing recalcitrant organics (pesticides), algal toxins, and soluble organic substances, which decrease the efficiency of chlorination toward viruses and other pathogens.

Examples of advanced wastewater treatment: South Tahoe and Pomona plants, California

At South Tahoe, California, an advanced wastewater treatment plant was constructed and later expanded from 2.5 to 7.5 mgd. This plant receives the effluent from an activated sludge unit and it includes the following steps:

1. *Lime flocculation.*
2. *Nitrogen removal tower.* (Ammonia stripping).

3. *Recarbonation step.* This decreases the pH from 11 to 9.3.
4. *Mixed-media filtration.* It uses coarse coal, sand, and fine garnet. Alum is added prior to filtration.
5. *Carbon adsorption.*
6. *Chlorination.* Final step.

This series of treatments produces a good quality effluent. Its characteristics are described in Table 6.7. A virus survey detected some viruses after the carbon column treatment, but none after the chlorination step. The efficiency of the chlorination step is partly due to the low turbidity of the effluent (0.3 JTU).

More recently, virus removal by a variety of advanced wastewater treatment systems, has been monitored at the Pomona research facility of the Los Angeles County Sanitation District. One of these systems, which include coagulation, sedimentation, filtration and disinfection (chlorine or ozone), is shown in Figure 6.17. The use of a sophisticated concentration method (improved version of the Carborandum virus concentrator) allowed the detection of 0.12 to 95 PFU/100 gal of advanced wastewater effluent. This survey also showed that

TABLE 6.7. Characteristics of Advanced Wastewater Effluent at South Tahoe, California[a]

Parameter	Level After Treatment
Turbidity	0.3 JTU
BOD	<1 ppm
COD	12 ppm
MBA S[b]	12 ppm
P	0.2 ppm
N	1–10 ppm
Coliforms	<2.2/100 ml
Virus[c]	Negative

[a] Adapted from Culp and Moyer (1969), *Civil Eng.* **39**:38–42.

[b] Measures synthetic detergents.

[c] Viruses were detected after activated carbon treatment but none after chlorination.

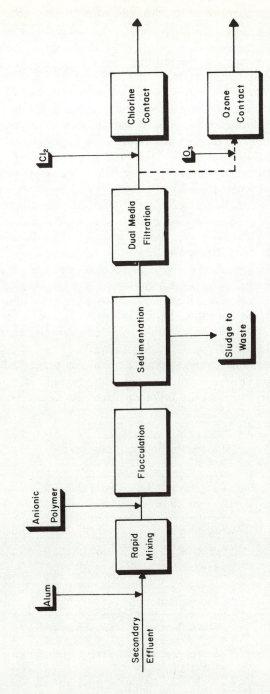

TERTIARY TREATMENT of WASTEWATER EFFLUENTS

FIG. 6.17. Tertiary treatment of wastewater effluents. Adapted from E. Garrison and R. P. Miele (1977), *J. Am. Water Works Assoc.* **69:**364–369.

the various tertiary treatment systems reduced the virus concentration by 4 to 6 $\log_{10}$.

The Fate of Viruses in Reclaimed Water

As world population increases, water will become a precious resource that will cost more and more to produce. Environmental engineers and scientists are now thinking about renovating wastewater to a degree suitable for recreational activities or even drinking purposes. At least for some countries, water demands will exceed freshwater supplies and future growth will depend on their ability to develop efficient wastewater reclamation plants. Such operations are undoubtedly costly, but they will be justified as the world demand for water increases. Wastewater renovation becomes even more attractive when one considers that 50 million people in the United States drink water of lower quality than that produced by advanced treatment (U.S. Public Health Service Survey of 1970). Moreover, raw sewage actually enters the drinking supply of many cities with a population over 25,000.

Following various degrees of treatments, wastewater is reused in 358 locations in the United States. Agriculture is the major user of reclaimed wastewater, which is also utilized for industrial (mostly cooling), recreational, and domestic (lawn irrigation, toilet flushing in motels, home laundry) purposes (Table 6.8). There are also some considerations given to the potential use of wastewater for fish propagation and farming.

In the United States, wastewater reclamation is particularly important for southern California as a supplement for water brought in from the northern part of the state and from the Colorado River. Reclaimed wastewater is used

TABLE 6.8. Municipal
Wastewater Reuse in the United
States[a]

Type of Reuse	1971 Volume (billions gal)
Irrigation	77
Industrial	53.5
Recreational	1.5
Domestic (nonpotable)	0.01

[a] From C. J. Schmidt, I. Kugelman, and E. V. Clements, III (1975), *J. Water Poll. Control Fed.* **47**:2229–2245. Courtesy of Water Poll. Contr. Fed.

TABLE 6.9.   Fate of Viruses, Coliphages, and Indicators Bacteria in the Stander (Pretoria, South Africa) Wastewater Reclamation Plant[a]

	Count/100 ml				
Treatment	Total Plate Count	Total Coliforms	Enterococci	Coliphages	Enteric Viruses
Raw Water	4.3×10^6	2.4×10^5	6.4×10^3	3.8×10^3	71
Lime Treatment[b]	1.8×10^4	33	186	1.4	1/21[e]
Quality Equalization[c]	3.3×10^4	10	81	1.7	1/14
Sand Filtration	6.3×10^5	163	2.4	2.1	1/25
Chlorination[d]	8×10^2	0	0	0	0/60
Activated Carbon	2.7×10^5	5	0	0	0/17

[a] Adapted from W. O. K. Grabow, B. W. Bateman, and J. S. Burger (1978), *Prog. Water Technol.* **10**:317–327.

[b] Lime treatment: The treatment lasts 50 minutes and the pH is 11.2.

[c] Quality equalization: Holding tank where the detention time is 10 hours.

[d] Chlorination: 2 mg/l of free available chlorine for a period of 2 hours.

[e] Number of positive sample/number of 10-liter samples tested.

for water recharge program (Whittier Narrows water reclamation), industrial reuse (cooling and process water), irrigation of agricultural land (Pomona plant), and recreational water reuse (Lancaster plant). In South Tahoe, California, reclaimed wastewater is transported through 17 miles of pipeline to Indian Creek reservoir, which is a man-made trout lake that can store up to a billion gallons of water. This lake is used for contact sports and for irrigation.

The subject of domestic use of reclaimed wastewater for potable purposes is still controversial and sometimes meets opposition from the public at large. One of the best examples of wastewater renovation for drinking and other purposes have been provided by researchers in South Africa. In 1969 the Windhoek (Namibia) wastewater reclamation plant was designed for the practice of large-scale reclamation of wastewater. Another plant, the Stander plant in Pretoria, was built for research purposes, and consists of the following unit processes:

1. *Lime treatment.* The pH is raised to 11.2.
2. *Quality equalization.* The water is held in tanks for a period of 10 hours.
3. *Sand filtration.*

4. *Chlorination.* The free available chlorine is 2 mg/l and the contact time is 2 hours.

5. *Activated carbon.*

6. *Final chlorination.* Reaches 1 mg/l of free available chlorine.

Table 6.9 shows the reduction of viruses, coliphages, and indicator bacteria in the Stander plant. It is clear that enteric viruses could not be detected in 10 liters effluent samples from that plant. Similar results have been reported at the Windhoek wastewater reclamation plant. It is believed that the total plate count is the most sensitive indicator of the quality of the water from these plants. Microbiological monitoring and epidemiological studies have shown that it is now technically possible to produce microbiologically safe water for domestic purposes. Unfortunately, other problems remain to be solved. The most pressing one is the fate of trace organics (e.g., chlorinated hydrocarbons) and their impact on human health. As far as the economic side is concerned there will come a time when the high cost of these operations will be justified.

6.4 AEROSOLS GENERATED BY SEWAGE TREATMENT PROCESSES: HEALTH IMPLICATIONS TO PLANT OPERATORS AND TO NEIGHBORING POPULATIONS

It has been known since the turn of the century that wastewater treatment processes assist in the aerosolization of microorganisms by air injection, spraying, or splashing. Once in the airborne state, microorganisms are subjected to meteorological conditions such as temperature, solar radiation, relative humidity, and atmospheric stability (this subject is examined in more detail in Chapter 9). Activated sludge and trickling filter units are the most important contributors to the production of biological aerosols. Airborne microbes have been detected up to almost 1 mile from sewage treatment plants. Most of the studies have dealt with the recovery of bacterial aerosols. However, there is a lack of information concerning the generation of viral aerosols by wastewater treatment plants. This is due to the fact that viruses occur in much lower concentrations than bacteria and to the lack of suitable recovery techniques. Nevertheless, many investigators have suggested that biological aerosols may cause respiratory and intestinal diseases in exposed sewage treatment plant operators. However, data gathered in the United States, Canada, and Europe do not show any significant increase in diseases following job-related exposure to microbial aerosols. It is possible that chronic exposure to low levels of pathogens may lead to an increased degree of immunity to airborne pathogens. It was found that for some enteroviruses (coxsackie B3), antibody levels were highest in wastewater treatment plant operators whereas for some others (echovirus 6) antibody levels were lower than in a control group. Nonetheless, despite the acquired immunity it is feared that these people may contribute to an increased incidence of enteric diseases among their immediate families. It is believed that nearby populations would be more susceptible because they are only sporadically exposed to bacterial and viral aerosols. There is still a lack of solid epidemiological evidence concerning the health risks of biological aerosols to plant workers and to nearby populations.

6.5   SUMMARY

1. Domestic wastewater harbors a wide range of bacterial and viral pathogens as well as protozoan and helminth parasites. These pathogens cause a wide range of diseases and other complications in humans.

2. In wastewater treatment plants, primary treatment does not remove significant amounts of viruses.

3. Among all the biological treatment processes, activated sludge is the most effective in virus reduction. The removal and/or inactivation of viruses is due to their adsorption to sludge solids and to their inactivation by sewage bacteria.

4. Virus removal by trickling filters is generally low and erratic.

5. In oxidation ponds physical (temperature, sunlight) and biological factors (algal–viral interaction) contribute to viral inactivation. Virus adsorption to solids and deposition in pond sediments play a significant role in virus removal.

6. Advanced treatment of domestic wastewater generally leads to a significant reduction in virus numbers.

7. Wastewater reclamation is becoming a reality. Well-designed wastewater reclamation plants are highly efficient in the removal of virus hazard from water.

8. There is no solid evidence to support that sewage treatment operators have a higher incidence of enteric diseases than control groups.

6.6   FURTHER READING

Berg. G., Ed. 1978. *Indicators of Viruses in Water and Food.* Ann Arbor Sci. Pub., Ann Arbor, Michigan.

Bitton, G. 1978. Survival of enteric viruses. In: *Water Pollution Microbiology*, Vol. 2. R. Mitchell, Ed. Wiley-Interscience, New York.

Clark, C. S., E. J. Cleary, G. M. Schiff, C. C. Linneman, Jr., J. P. Phair, and T. M. Briggs. 1976. Disease risks of occupational exposure to sewage. *J. Environ. Eng. Div.* **102**:375–388.

Geldreich, E. E. 1972. Water-borne pathogens. In: *Water Pollution Microbiology.* R. Mitchell, Ed. Wiley-Interscience, New York.

Hays, B. D. 1977. Potential for parasitic disease transmission with land applications of sewage plant effluents and sludges. *Water Res.* **11**:583–595.

Malina, J. F., Jr. 1976. Viral pathogens inactivation during treatment of municipal wastewater. In: *Virus Aspects of Applying Municipal Wastes to Land.* L. B. Baldwin et al., Eds. University of Florida, Gainesville, Fla.

Miele, R. P., and W. E. Garrison. 1977. Current trends in water reclamation technology. J. AWWA. **69**:364–369.

Mitchell, R. 1974. *Introduction to Environmental Microbiology.* Prentice-Hall, Englewood Cliffs, N.J.

Pikes, E. B., and C. R. Curds. 1971. The microbial ecology of the activated sludge process. In: *Microbial Aspects of Pollution*, G. Sykes and F. A. Skinner, Eds. Academic, New York.

Zajic, J. E. 1971. *Water Pollution: Disposal and Re-use*, Vol. 1. Marcel Dekker, New York.

seven
Fate of Viruses in Water Treatment Plants

The quest for pure water has been pursued for at least 4000 years. The search is not ended. More than likely, it is never ending because new materials, new industrial processes, new attitudes and preferences, and new criteria will continue to arise. They will have to be eternally assessed and value judgements be assigned.

ABEL WOLMAN

(1970), *J. Am. Water Works Assoc.* **62**:746

7.1 INTRODUCTION

Rising populations in large metropolitan areas have resulted in an ever-increasing demand for potable water. Groundwater supplies have become insufficient in filling our need for safe water. Communities are now turning to surface waters as a source of potable water. In the United States surface waters constitute the major source of water supply for communities of 25,000 people or more. This is illustrated in Table 7.1. It is shown that 53% of the communities surveyed derive their drinking water from surface waters, as compared to 28% from groundwater. However, surface waters are commonly polluted by industrial and municipal sewage that contains a wide range of microbial pathogens, including viruses. In some areas, these surface waters are practically "diluted wastewaters." Thus, viral contamination of water supplies is real and has been documented in various parts of the globe (Table 7.2). Problems arise when upstream communities discharge virus-laden sewage into surface waters that become drinking water supplies to downstream communities. Undoubtedly water treatment operations do not always insure a complete removal or inactivation of viruses. From France came the first report on virus isolation from treated municipal water: Up to 19% of samples collected at the distribution pipes in Paris were positive for virus. This "disturbing" report triggered a worldwide search for the presence of these infectious agents in drinking waters. An examination of water treatment plants in Nancy and Lunéville in France, revealed the presence of coxsackie B virus after coagulation and sand filtration but none after disinfection with chlorine or ozone. A virological monitoring, carried out in Cincinnati, Ohio, during an 11-month period (1963–1964), did not reveal the presence of any virus in treated drinking water. Then in 1970 viruses were isolated from drinking water from two communities in Massachusetts. This led the EPA's Water Supply Research Laboratory, in Cincinnati, to conduct a study on viruses in drinking water. Three EPA laboratories

TABLE 7.1.   Source of Water Supply for U.S. Communities of 25,000 People or More[a]

	Communities	
Supply Source	Number	% of Total
Surface	414	53.0
Ground	219	28.0
Surface + ground	148	19.0

[a] Adapted from G. F. Craun and L. J. McCabe (1973), *J. Am. Water Works Assoc.* **65:**74–84.

TABLE 7.2. Global Aspect of Virus Contamination of Water Supplies[a]

Type of Water	% Samples Positive for Enteroviruses
River water (France)	21; 9
River water (USSR)	34
River water (Switzerland)	38; 63
River water (Illinois River, USA)	27
Domestic water supply (Israel)	2
Domestic water supply (England)	56

[a] Adapted from J. T. Cookson (1976), *J. Am. Water Works Assoc.* **66:**707–711.

participated in the virus monitoring program: Gulf Coast, Northeast, and Northwest laboratories. Their findings are summarized in Table 7.3. No virus was detected by the Gulf coast and Northwest laboratories. However one poliovirus type 3 was found by the Northeast laboratory. It was suggested that this virus was probably the result of laboratory contamination. In this type of study

TABLE 7.3. Monitoring of Drinking Water for Viruses[a]

No. Samples	Volume of Sample (Gal)	Free Chlorine (mg/L)	pH Range	Turbidity	Virus Recovered
		Gulf Coast Laboratory Study			
25	129	0.85	7.5–9.4	0.03	None
18	111	2.10	8.1–9.5	0.40	None
		Northwest Laboratory Study			
6	100	0.29	7.9–9.0	0.23	None
6	100	0.28	7.9–8.6	0.26	None
		Northeast Laboratory Study			
7	2–5	0.81	7.2–9.1	0.10	Polio 3
4	4–250	1.11	6.9–8.2	0.29	None

[a] Adapted from N. A. Clarke et al. (1975), *J. Am. Water Works Assoc.* **67:**192–197.

genetic markers (see Chapter 1) that differentiate between laboratory and wild-type polioviruses need to be examined.

It can be argued that the inability to detect viruses is probably due to the low efficiency of the concentration method (i.e., gauze pad method) used in many of the previous studies. A recent investigation, using a more sophisticated concentration technique, has revealed the presence of enteroviruses on four occasions in finished waters in the Occoguan Reservoir area, near Washington, D.C. However EPA could not confirm the presence of enteric viruses in these waters. It remains to be seen if the use of these improved concentration techniques will enable us to recover viruses from drinking water in other areas.

The problem of viruses in drinking water can be better understood only after examination of the processes involved in water treatment facilities and their impact on virus removal or inactivation.

7.2 DESCRIPTION OF A WATER TREATMENT PLANT

Water used for domestic purposes (drinking and food preparation) must be *safe* and *aesthetically* acceptable to the consumer. To attain this goal water treatment plants must remove efficiently biological and chemical pollutants that may constitute a health hazard to man. They must also produce a water free from turbidity, color, taste, and odor. Thus in a treatment plant raw water undergoes a series of physicochemical processes, some of which have been known for centuries.

Two categories of water treatment plants are discussed in the following sections.

7.2.1 Conventional Filter Plant

This type of plant (Figure 7.1) is based on the processes of coagulation and filtration. It includes the following steps:

1. *Mixing basin.* The raw water is rapidly mixed with the coagulant.
2. *Flocculating basin.* After the coagulation process, an appropriate mixing will produce large flocs.

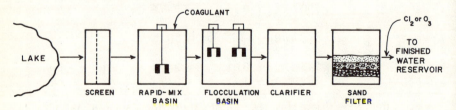

FIG. 7.1.   Flow diagram of a conventional filtration plant.

3. *Clarifier.* It is a basin where the flocs settle out.

4. *Filtration.* The effluent from the clarifier is passed through a rapid sand filter or a diatomaceous earth filter.

5. *Disinfection with a strong oxidant.* This topic will be covered in detail in Chapter 8.

7.2.2   Softening Plants

These plants are designed for the treatment of *hard* water (water high in Ca and Mg compounds). Their operation is based on four basic processes:

1. *Softening process.* This process removes the water hardness and results in the formation of precipitates of Ca and Mg.

2. *Settling of the precipitates.* Sometimes a coagulant aid may be used to help the settling process.

3. *Filtration.*

4. *Disinfection.*

Thus, in water treatment plants, viruses and other microorganisms may be physically removed by processes such as coagulation, precipitation, filtration, and adsorption, or inactivated (i.e., killed) by disinfection, or by high pHs as a result of softening.

We will now describe in more detail each one of these processes and discuss their effect on the fate of pathogenic viruses.

7.3   FATE OF VIRUSES FOLLOWING WATER TREATMENT OPERATIONS

7.3.1   Storage

Prior to treatment water may be stored in natural or artificial basins. During storage viruses are subjected to the whole range of environmental factors that are examined in Chapters 3 and 4. Temperature is the most significant factor controlling virus survival in storage basins. Since some viruses are slowly inactivated, it can be concluded that storage does not contribute a great deal to their reduction in water treatment plants.

7.3.2   Clarification

Clarification is the process that removes turbidity and color from water. A well-designed clarification process followed by filtration results in a clear and low color effluent with 0.1 JTU and three color units as recommended by the

American Water Works Association (AWWA). Since turbidity is predominantly made of colloids, plain sedimentation is not sufficient for the settling of the fine colloidal particles. One thus turns to the process of *coagulation* by which colloidal particles agglomerate and produce aggregates that are large enough to settle out rapidly. Coagulation is the most important process used in water treatment plants for the clarification of colored and turbid waters.

Coagulation involves the destablization of colloidal particles (e.g., bacteria, viruses) by coagulants (Al and Fe salts) and sometimes by coagulant aids (bentonite, polyelectrolytes, starch, activated silica). The most commonly used coagulants are:

- Aluminum sulfate: $Al_2(SO_4)_3 \cdot 18H_2O$ (Alum).
- Ferric sulfate: $Fe_2(SO_4)_3$.
- Ferric chloride: $FeCl_3$.

It is now accepted that the coagulation process involves an interaction between negatively charged hydrophobic or hydrophilic colloidal particles (clays, bacteria, viruses, etc.), and positively charged hydrolysis products of Al and Fe. Many factors influence the coagulation process. The most important one is the pH of the water. The optimum coagulation with alum lies between pH 6 and 7.5, whereas with iron the optimum pH range is broader. Other factors include turbidity, temperature, and mixing. When the optimal conditions are met it appears that coagulation–sedimentation contribute significantly to the removal of viruses in water treatment plants. Alum, at a concentration of 25 to 100 mg/l removes from 98 to over 99% of the virus present in water (Table 7.4). Ferric chloride and ferric sulfate, under appropriate concentration and conditions, perform as well as alum. It is essential to note that the removal of viruses under field situation is generally lower than in laboratory controlled conditions. In field situations, the flocculation process is influenced by many uncontrolable factors.

The coagulation process does not inactivate viruses. There is a mere transfer of pathogenic viruses from water to the flocculated material. It is thus important to dispose of this material in a proper way. Coagulation contributes indirectly to the inactivation or removal of viruses:

1. By removing organic matter, turbidity and color, and thus making the disinfection process more effective.
2. Coagulation is generally followed by sand filtration and this combination is effective in removing turbidity in general and viruses in particular.

It is sometimes necessary to improve the coagulation process by using a *coagulant aid* such as bentonite (10–50 ppm), activated silica (7–11% of the alum

TABLE 7.4.   Removal of Viruses by Coagulation–Sedimentation[a]

Coagulant	Concentration of Coagulant (mg/l)	Clay (mg/l)	Virus	Removals Virus (%)	Turbidity (%)
Alum	10	50	Poliovirus 1	86	96
	25.7	120	Phage T4	98	99
	25.7	120	Phage MS2	99.8	98
Ferric Sulfate	40	—	Poliovirus 1	99.8	—
	40	—	Phage T4	99.8	—
Ferric chloride	60	50	Poliovirus 1	97.8	97.5[b]
	60	100	Poliovirus 1	93.3	97.8[b]
	60	500	Poliovirus 1	99.7	99.7[b]

[a] Data adapted from M. Chaudhuri and R. S. Engelbrecht (1970), *J. Am. Water Works Assoc.* **62**:563; J. M. Foliguet and F. Doncoeur (1975), *Water Res.* **9**:953; Guy et al. (1977), *Water Res.* **11**:421; R. T. Thorup et al. (1970), *J. Am. Water Works Assoc.* **62**:97.
[b] Turbidity removal after sand filtration.

dose), or polyectrolytes (0.1–1 ppm). The coagulant aid increases the floc size and thus leads to a faster settling rate. The most promising coagulant aids are polyelectrolytes, which are polymers of synthetic origin. On the basis of their ionogenic charges, they are classified as cationic, anionic, and noionic poly-electrolytes. The concentration of polyelectrolytes is an important factor governing their action. At optimum concentrations polyelectrolytes form flocs by "bridging" between particles. An excess amount of these polymers will redisperse the system (Figure 7.2). This phenomenon characterizes both the behavior of biological polymers (polysaccharides) and synthetic polyelectrolytes.

Under controlled conditions polyelectrolytes are able to interact with viruses. Cationic polyelectrolytes generally display a higher sorptive capacity than nonionic or anionic polyelectrolytes. This is explained by the possible interaction between positively charges anino groups on the polyelectrolyte surface and negatively charged virus particles.

Since some polyelectrolytes display a high affinity toward viruses it was thought that their use as coagulant aids may improve the coagulation process. Anionic and nonionic polyelectrolytes do not affect or may even decrease the efficiency of coagulation in virus removal. Cationic polyelectrolytes slightly improve virus removal (Table 7.5).

However, as in the case of primary coagulants, polyelectrolytes may indirectly increase virus removal in processes such as sand or diatomaceous earth filtration. They produce strong flocs that resist shear forces that may pre-

DESTABILIZATION OF COLLOIDAL PARTICLES
BY A POLYELECTROLYTE FLOCCULANT

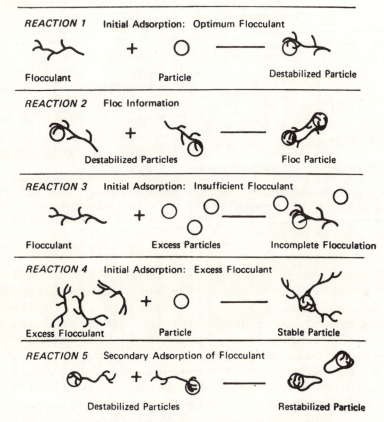

FIG. 7.2. Destabilization of colloidal particles by a polyectrolyte flocculant. From S. L. Daniels (1980), in: *Adsorption of Microorganisms to Surfaces*, G. Bitton and K. C. Marshall, Eds., Wiley, New York.

vail during sand filtration. Virus removal is also significantly enhanced in polyelectrolyte-coated diatomite filters.

7.3.3   Filtration

Filtration is defined as the passage of fluids through a porous medium to remove suspended solids such as clays, silt particles, microbial cells, and flocculated material. This process depends on the filter medium, concentration and type of solids to be filtered out, and the operation of the filter.

Filtration may involve:

- The passage of water through a filter aid which has accumulated on a support septum. This is essentially the principle behind diatomaceous earth filtration.

- The downward flow of water through a bed of filter medium. The function of *rapid sand filters* is based on this principle.

We will now get acquainted with these two filtration processes and discuss their role in virus removal.

Diatomaceous Earth Filtration

Diatomite filters were used by the army during World War II to remove amoebic cysts from canteen water. Later they were employed in swimming pools and in municipal water treatment plants.

Water containing diatomite is passed through a porous *septum* which acts as a support medium for the buildup of an ⅛-in. *precoat* of filter medium (Figure 7.3). The filter is now ready for operation. Raw water is filtered at a flow rate of 0.5 to 2 gpm/ft² through the precoat. During filtration, diatomite may be continuously added to keep the filter running for longer periods of time. This additional diatomite is called *body feed*.

Diatomite filters are made of remains of siliceous shells of diatoms and are supplied in a variety of grades. The septum is cleaned following a certain period of operation and the filter cake should be disposed of properly.

TABLE 7.5.   Efficacy of Polyelectrolytes (Coagulant Aids) in the Removal of Poliovirus Type 1 by Coagulation–Sedimentation[a,b]

Polyelectrolyte	Polyelectrolyte Concentration (ppm)	Virus Removal (%)
No polyelectrolyte		97.81
Cationic polyelectrolyte (C33)	0.25–1	99.72
Anionic polyelectrolyte (AP22)	0.25 1	88.82 95.83
Nonionic polyelectrolyte (NIT)	0.25	92.55

[a] Adapted from J. M. Foliguet and F. Doncoeur (1975), *Water Res.* **9**:953–961.
[b] Coagulation with FeCl₃ at a concentration of 60 mg/l.

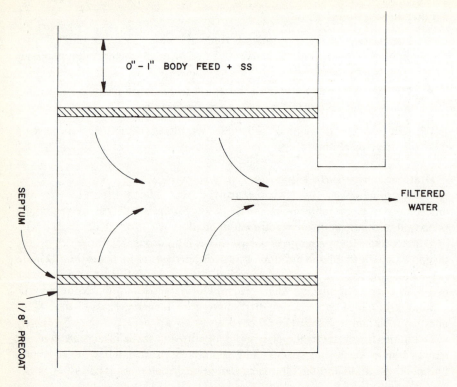

FIG. 7.3.   Diatomite filter. From E. R. Baumann (1971), in: *Water Quality and Treatment* (AWWA), McGraw-Hill Book Co., New York. Courtesy of American Water Works Association.

The removal of viruses by diatomaceous earth filters is generally very poor. The removal may be enhanced by coating the filter medium with iron and aluminum salts or with cationic polyelectrolytes. Table 7.6 shows the removal of poliovirus by diatomaceous earth coated with various chemicals. A cationic polyelectrolyte at a concentration of 0.15 ppm brings about the highest removal of poliovirus. Polyelectrolytes may also be added in the suspending medium in lieu of coating. In that case their concentration in the water is critical. Other factors govern virus removal by diatomite filtration. These include the level of coating of the filter, pH of the suspending medium, and virus level.

Rapid Sand Filtration
A rapid sand filter consists of a concrete box filled with 24 to 30 in. of sand (0.5 mm effective diameter) supported by a layer of anthracite, gravel, or calcite (Figure 7.4). An underdrain system collects the water which has filtered down-

TABLE 7.6.   Removal of Poliovirus by Diatomaceous Earth Filtration[a]

Filter Aid	% Removal (after 2 hours of operation)
Hyflo	62
Hyflo coated with 2% aluminum hydrate	76
Hyflo coated with 2% ferric hydrate	98
Hyflo + cationic polyelectrolyte (0.15 ppm)	100

[a] Adapted from T. S. Brown et al. (1974), *in: Virus Survival in Water and Wastewater Systems,* J. F. Malina, Jr. and B. P. Sagik, Eds., University of Texas, Austin.

ward through the sand. The flow rate through such a filter varies from 2 to 4 gpm/ft². To maintain effectiveness, the sand filter must be cleaned by reversing the flow. This backwashing is done at a sufficient flow rate to allow a thorough cleaning of the sand. A sand filter is practically always used to process water that has been previously treated by coagulation or by lime-soda softening.

Sand particles are essentially poor adsorbents toward viruses. Sand–virus interaction is governed by the physicochemical properties of the suspending medium, the size of the sand particles, and by the flow rate. Adsorption and filtration are the two processes that contribute to the removal of viruses by sand filters. For instance viruses attached to solid particles or to flocs are filtered out of the water when passing through a sand filter. In general the removal of viruses by sand filtration is variable and often low. However, coagulation of the virus prior to sand filtration removes more than 99% of viruses (Table 7.7).

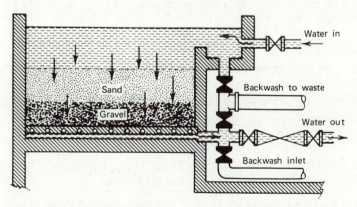

FIG. 7.4.   Diagram of a sand filter. From S. D. Strauss (1973), *Water Treatment* (A *Power* Special Report). Courtesy of *Power.*

TABLE 7.7.   Removal of Poliovirus by Rapid Sand Filters[a,b]

Treatment	% Removal
A. Sand filtration	1–50
B. Sand filtration + coagulation with alum	
1. without settling	90–99
2. with settling	>99.7

[a] Adapted from G. Robeck et al. (1962), *J. Am. Water Works Assoc.*
54:1275–1290.
[b] The flow rate was 2–6 gpm/ft².

The performance of sand filters may also depend on the frequency of backwashing. A recently backwashed sand filter may not be very effective in retaining viral particles.

7.3.4   Water Softening

Hardness is essentially due to the presence of calcium and magnesium compounds in the water. There are two categories of hardness:

- *Carbonate hardness.* It is caused by bicarbonates of Ca and Mg.
- *Noncarbonate hardness.* It is due to chlorides and sulfates of Ca and Mg.

The concept of hardness has practical implications, since the use of hard water for domestic purposes necessitates a higher consumption of soap. It is sometimes necessary to *soften* the water, i.e., to remove Ca and Mg compounds. This can be done by using one of two processes: lime-soda process or ion-exchange resins.

Lime–Soda Process

In this process, the softening chemicals are lime (calcium hydroxyde) and soda ash (sodium carbonate).

The removal of carbonate hardness follows the equations:

$$Ca(HCO_3)_2 + Ca(OH)_2 \rightarrow 2CaCO_3\downarrow + 2H_2O \tag{7.1}$$
$$\text{Ca-bicarbonate + lime}\qquad \text{Ca-carbonate}$$

$$Mg(HCO_3)_2 + 2Ca(OH)_2 \rightarrow Mg(OH)_2\downarrow + CaCO_3 + 2H_2O \tag{7.2}$$
$$\text{Mg-bicarbonate + lime}\qquad \text{Mg-hydroxide}$$

From the above equations it is clear that the addition of lime results in the formation of precipitates of $CaCO_3$ (calcium carbonate) and/or $Mg(OH)_2$ (magnesium hydroxide).

Sodium carbonate (Na_2CO_3) is used for the removal of noncarbonate hardness.

$$CaSO_4 + Na_2CO_3 \rightarrow CaCO_3\downarrow + Na_2SO_4 \tag{7.3}$$

$$MgSO_4 + Ca(OH)_2 + Na_2CO_3 \rightarrow Mg(OH)_2\downarrow + CaCO_3\downarrow + Na_2SO_4 \tag{7.4}$$

Equations 7.2 and 7.4 show that there are additional costs involved in the removal of Mg hardness. After looking at the chemical basis of the lime-soda process, we will now discuss the fate of viruses following this particular water treatment process.

Table 7.8 shows the removal of poliovirus following lime or lime-soda ash softening. The precipitation of only $CaCO_3$ does not result in a significant virus removal. However, the lime-soda ash softening, which leads to the precipitation of both $CaCO_3$ and $Mg(OH)_2$, results in a much higher virus removal (>99.9%). The high virus removal obtained by using magnesium hydroxide may be explained by electrostatic attraction between negatively charged viruses and positively charged $Mg(OH)_2$. Calcium carbonate is negatively charged and should have less affinity toward virus particles (Figure 7.5).

As seen in Table 7.8, the lime-soda ash process is also associated with pH values above 11. This high pH does contribute to the virus inactivation by denaturing their protein coat.

Thus, the lime-soda ash process results in a *physical removal of viruses* by adsorption to magnesium hydroxide flocs and in their *inactivation* by the high pH generated by the process.

TABLE 7.8.   Removal of Viruses by Water Softening Process[a]

Initial Conditions Hardness as $CaCO_3$ (mg/l)		Final Conditions			
		Hardness as $CaCO_3$			% Removal of
$MgCl_2$	$Ca(HCO_3)_2$	Mg	Ca	pH	Poliovirus
Lime Softening					
—	100	—	64	9.0	9
—	300	—	218	8.1	70
Lime-Soda Ash Process					
67	133	50	76	10.8	99.90
100	300	82	204	11.2	99.993

[a] Adapted from D. F. Wentworth et al. (1968), *J. Am. Water Works Assoc.* **60**:939.

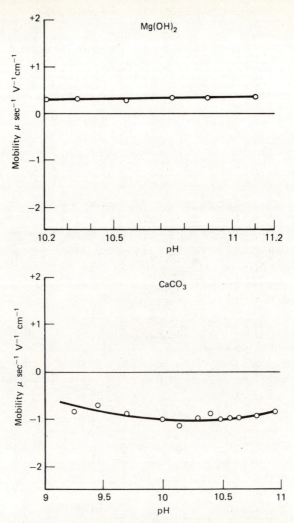

FIG. 7.5. Electrophoretic mobility of calcium carbonate and magnesium hydroxide flocs. From A. P. Black and R. F. Christman. (1961), *J. Am. Water Works Assoc.* **53**:737–747. Courtesy of American Water Works Association.

Ion-Exchange Resins

An ion-exchange resin is a synthetic matrix of styrene divinyl copolymer which has the ability to exchange one ion for another. It can be viewed as a large polymeric network which assumes either a positive $[R—N^+(CH_3)_3$ for anion-exchange resins] or negative ($R—SO_3^-$ for cation-exchange resin) charge in association with counter ions of opposite charge.

In water softening we are interested in exchanging Ca and Mg for Na. For that purpose, we use cation exchange resins saturated with Na^+. Once the *cation-exchange capacity* of the resin is exhausted, it is regenerated with a salt, NaCl. A typical ion exchange unit used in water softening plants is shown in Figure 7.6.

The following equations show the exchange reactions involved in water softening:

- *Carbonate hardness:*

$$Ca(HCO_3)_2 + Na_2R \rightarrow CaR + 2NaHCO_3 \qquad (7.5)$$

$$Mg(HCO_3)_2 + Na_2R \rightarrow MgR + 2NaHCO_3 \qquad (7.6)$$

- *Noncarbonate hardness:*

$$CaSO_4 + Na_2R \rightarrow CaR + Na_2SO_4 \qquad (7.7)$$

$$MgCl_2 + Na_2R \rightarrow MgR + 2NaCl \qquad (7.8)$$

In both cases (carbonate and noncarbonate hardness) Ca and Mg are removed from water by exchange with Na present on the exchange resin sites.

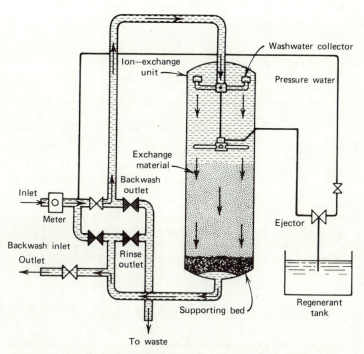

FIG. 7.6. A typical ion-exchange resin. From S. D. Strauss (1973), *Water Treatment* (A *Power* Special Report). Courtesy of *Power*.

The removal of viruses by ion-exchange resins has been known for about two decades. Essentially viruses are well removed by anion-exchange resins but are not well retained by cation-exchange resins. The latter bind viruses only in the presence of enough cations to neutralize the negative charges on both the resin and the virus.

7.3.5   Activated Carbon

Activated carbon originates from various sources such as wood, bituminous coal, lignite, and petroleum base residues. It is activated by a combustion process. It offers a large surface ($500-600$ m^2/g) for the adsorption of tastes and odors, excess chlorine, and organic materials in general. It is supplied in granular or powdered form. Beds of granular activated carbon can be used at any stage in the water treatment plant.

As in the case of other waste treatment processes, activated carbon has been investigated for its adsorptive capacity toward viruses. These studies have shown that viruses are adsorbed to activated carbon by electrostatic attraction between positively charged amino groups (NH_3^+) on the virus and negatively charged carboxyl groups (COO^-) on the surface of the carbon. This interaction is controlled by *pH, ionic strength,* and *organic matter content* of the suspending medium. Activated carbon does not remove significant amounts of viruses

TABLE 7.9.   Virus Removal by Activated Caron in Water Treatment Plants[a]

Virus	Load (Infective Particles)	Removal (%)	Days After Backwashing
Attenuated poliovirus	1.2×10^4	75.0	>7
	3.5×10^3	82.0	>7
Phage T4	9.6×10^8	81.3	>7
	5.1×10^8	78.4	>7
	3.5×10^8	74.9	>7
	5.9×10^7	76.3	>7
	4.0×10^7	>86.0	>7
	1.5×10^7	>64.0	>7
	1.4×10^9	21.4	2
	2.1×10^7	47.6	2
	2.1×10^8	53.3	3

[a] Adapted from M. D. Guy, et al. (1977), *Water Res.* **11**:421–428.

from water. The low removal results from heavy competition of organics with viruses for the attachment sites on the carbon surface.

Table 7.9 illustrates the performance of activated carbon in the removal of added attenuated poliovirus and bacteriophage T4 from Trent River water (England). After 7 days without backwashing, this treatment removed 75 to 82% of poliovirus and from 64 to 86% of phage T4. It also seems that the efficiency of activated carbon treatment is dependent on the frequency of backwashing.

7.4   VIRUS REMOVAL BY PILOT WATER TREATMENT PLANTS

Studies in England have shown that a pilot water treatment plant was very effective in virus removal, which was greater than 99.9995%. A schematic representation of this pilot plant is shown in Figure 7.7. Among the various stages involved (flocculation–sedimentation, rapid sand filtration, activated carbon treatment, chlorination), the flocculation–sedimentation step removed the majority of viruses added to the Trent river water.

7.5   HEALTH ASPECTS OF CHEMICAL POLLUTANTS ASSOCIATED WITH DRINKING WATER SUPPLIES

We have discussed the fate of pathogenic viruses in water treatment plants. Chemical pollutants present in water supplies are of no less concern and their detection and removal are a new challenge to water treatment experts. As our analytical techniques become more sophisticated, a growing number of these chemical pollutants is detected. However, their impact on human health has not been fully assessed. These chemicals include nitrates, heavy metals, polynuclear aromatic hydrocarbons, halogenated organics, and asbestos fibers, to cite a few. Some of the trace organics are present in minute quantities in water and are potentially carcinogenic to humans. For example, polynuclear aromatic hydrocarbons, mostly benzo(a)pyrene, have been found in finished drinking waters, but their impact on human health is not fully understood. Moreover, chlorination of natural waters results in the production of chlorinated hydrocarbons, mainly chloroform ($CHCl_3$). Some of these chemicals are known to be carcinogenic and serious concerns have been raised over their long-term effects on our health.

Human exposure to various chemicals has been the subject of much speculation. It was observed that certain chemicals have an effect on virus infectivity. Commercial emulsifiers, especially the nonionic polymers of polyoxyethylene ethers, are capable of enhancing the sensitivity of cultured mammalian cells to

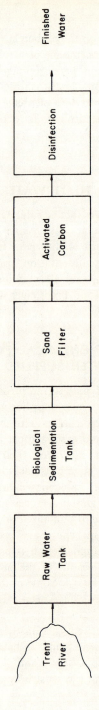

FIG. 7.7. A schematic representation of a pilot plant established by the Water Research Association in England. (Adapted from M. D. Guy et al.(1977), *Water Res.* **11**:421–428.

infection with several single-stranded RNA viruses such as vesicular stomatitis (VSV) virus, encephalomyocarditis (EMC) virus and poliovirus type 1.

7.6 DETECTION OF VIRUSES IN DRINKING WATER

We have discussed in detail the virus concentration methodology in Chapter 5. Since the number of viruses present in tapwater may be very low, one must be able to process at least 100 or even 1000 gallons of tapwater in order to detect some plaque-forming units. The concentration techniques that have been recently developed are based on the process of adsorption of viruses to a membrane filter followed by elution with an alkaline proteinaceous material. Figure 7.8 shows a virus concentrator developed by EPA researchers in Cincinnati, Ohio. This apparatus consists essentially of a flow meter, a pressure gauge, a fluid proportioner with two pumps that dose the chemical additives (HCl and AlCl$_3$), a mixing chamber and finally a filter holder. The 14th edition of *Standard Methods for the Examination of Water and Wastewater* contains a tentative virus concentration technique which makes use of such an apparatus. The method has a recovery efficiency of approximately 35% for highly polished tapwater and the detection of 3–7 PFU/100 gallons has been reported. Another group of researchers at Baylor College of Medicine in Houston, Texas used pleated membrane filters to retain viruses. These cartridge filters have a high surface area and process water at flow rates in the order of 5 to 10 gal/min. The recovery efficiency is approximately 40 to 50%. Other methods which use eluents at pH 9.0 (casein, beef extract, lysine) have been discussed in Chapter 5.

One may conclude that we now have the technological capability for the detection of low numbers of viruses in large volumes of tapwater. This knowledge is essential for the establishment of a virus standard for drinking water. However, some questions remain unanswered: cost of the method and trained personnel to handle this delicate task. One must not forget that in our technological society, water is still one of the cheapest resources available. Therefore, the consumer may have to pay an extra cost for the best water treatment and monitoring if he wants safe water in his kitchen and bathroom. Increased attention is being paid to the question of dual water systems for household and other urban uses. The dual water system will produce two grades of water, one for drinking purposes, and the other for miscellaneous uses. This is a sound idea since only 5% of the total water used by American consumers (160 gal water per person per day) is intended for drinking and cooking. The economic feasibility of this system is now under consideration in studies carried out in Tampa–St. Petersburg and West Philadelphia.

Water used for drinking purposes must be free from biological contaminants (viruses, bacteria, amoebic cysts) and this goal requires an adequate treatment

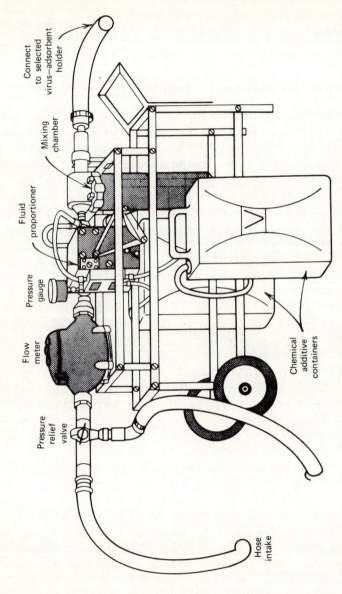

FIG. 7.8. Concentrator used for virus detection in drinking water. From W. F. Hill et al. (1974), *Appl. Microbiol.* **27**:1176–1177. Courtesy of American Society for Microbiology.

by softening, coagulation, filtration and disinfection with chlorine or ozone. Following passage through a well-designed water treatment plant, viruses will be reduced by 3 to more than 6 orders of magnitude.

However, viruses may also be expected to pass through conventional water treatment plants, especially through those that are poorly operated and where disinfection is inadequate. It is worth noting that the presence of viruses in drinking water has been demonstrated in many countries, particularly in India (1–7 PFU/12–40 liters!), United States, U.S.S.R., France, South Africa, and Romania. Furthermore, nontreated well-water may contain viruses and may be the cause of epidemics of viral origin (e.g., recent epidemic of hepatitis A in China).

7.7 SUMMARY

Among the various stages involved in a water treatment plant, the coagulation–sedimentation step is the most effective in virus removal. Cationic polyelectrolytes used as coagulant aids may slightly increase virus removal by coagulation.

In water softening plants the lime-soda ash process results in a removal of viruses by adsorption to magnesium hydroxide flocs and in their inactivation by the high pH generated by the process. Activated carbon treatment does not adequately remove viruses from water.

Pilot plant studies have shown that in a well-designed water treatment plant virus reduction may be greater than 99.9995%.

There are now adequate techniques for the detection of low numbers of viruses in large volumes of drinking water. However, their efficiency has been demonstrated only for the enterovirus group.

7.8 FURTHER READING

American Water Works Association. 1971. *Water Quality and Treatment.* McGraw-Hill, New York.

Baker, M. N. 1969. *The Quest for Water.* American Water Works Association, New York.

Berg, G. 1971. Removal of virus from water and wastewater. In: 13th water quality conference: "*Virus and Water Quality: Occurrence and Control.* University of Illinois at Urbana–Champaigne.

Bitton, G. 1975. Adsorption of viruses onto surfaces in soil and water. *Water Res.* **9:**473–484.

Bitton, G., and K. C. Marshall, Eds. (1980). *Adsorption of Microorganisms to Surfaces.* Wiley, New York.

Committee of Environmental Quality Management. 1970. Engineering evaluation of virus hazards in water. J. San. Eng. Div. **96:**111–169.

Schwartzbrod, L., J. Schwartzbrod, F. Doncoeur, J. M. Foliguet, and F. Lautier. 1973. Virus et eaux: I. Enterovirus et traitement des eaux usées et des eaux destinées à l'alimentation humaine. *Rev. Epidemiol. Med. Soc. Santé Publ.* **21:**99–116.

Sobsey, M. D. 1975. Enteric viruses and drinking water supplies. *J. Am. Water Works Assoc.* **67:**414–418.

Sproul, O. J. 1976. Removal of viruses by treatment processes. In: *Viruses in Water*. G. Berg et al., Ed. American Public Health Association, Washington, D.C.

World Health Organization, 1979. WHO Scientific group on Human Viruses in Water, Wastewater and Soil. Geneva, 23–27 October 1978.

eight
Virus Inactivation by Disinfectants in Water and Wastewater

8.1 INTRODUCTION

Disinfection is a process that is aimed at the destruction of all microorganisms capable of producing disease. A disinfected water is not, however, necessarily sterile.

Disinfection of water and wastewater can be accomplished by using chemical (halogens, ozone) or physical (ultraviolet irradiation, photodynamic inactivation) means. For many years chlorine has been the major disinfectant used in water and wastewater treatment plants. During the last few years chlorination has been challenged due to the development of major events that prompted a reassessment of the whole concept of disinfection. These events include the following:

1. The development of improved detection techniques for pathogens, particularly viruses. The use of these techniques has revealed that chlorine does not completely inactivate viruses in water and wastewater.

2. The passage of the Safe Drinking Water Act of 1976, which resulted in more rigid standards.

3. The discovery of chlorinated hydrocarbons in water supplies and in finished drinking water. These compounds (e.g., chloroform, carbon tetrachloride) are suspected carcinogens and are produced by the reaction of chlorine with trace organics present in water.

These events have prompted a search for alternatives to chlorination and for means to remove the precursor trace organics that react with chlorine.

In this chapter, we will briefly review the chemistry of halogens and ozone, and examine chemical and physical means for virus destruction in water and wastewater.

8.2   HALOGEN AND OZONE CHEMISTRY

We have already seen that water treatment processes such as coagulation, softening or sand filtration contribute to the removal of pathogenic viruses. Disinfection is undoubtedly the last line of defense against microbial pathogens, and halogens (chlorine, iodine, bromine) are used for that purpose. Among the halogens used in water and wastewater treatment chlorine is still the most widely employed chemical and its chemistry in water will be examined in some detail.

8.2.1   Chlorine

The use of chlorine as a water disinfectant goes back as far as the beginning of the nineteenth century. It was not until the advent of the 20th century that chlorine was used for the disinfection of domestic water supplies.

If chlorine gas (Cl_2) is introduced in water, it will hydrolize following the equation:

$$Cl_2 + H_2O \rightleftarrows HOCl + H^+ + Cl^- \qquad (8.1)$$

$\uparrow$ $\qquad\qquad$ $\uparrow$

Chlorine $\qquad$ Hypochlorous

Gas $\qquad\qquad$ Acid

The hypochlorous acid dissociates in water according to pH and temperature conditions:

$$HOCl \rightleftarrows H^+ + OCl^- \qquad (8.2)$$

$\uparrow$ $\qquad\qquad$ $\uparrow$

Hypochlorous $\quad$ Hypochlorite

Acid $\qquad\qquad$ Ion

The proportion of HOCl and OCl^- greatly depends on the pH of the water. Figure 8.1 shows that, at pH < 6, hypochlorous acid predominates whereas, at pH > 9.5, hypochlorite is the prevalent form. Chlorine present in water as HOCl or OCl^- is defined as *free available chlorine*.

When chlorine salts (sodium hypochlorite, calcium hypochlorite) are used, their dissociation in water will take place according to Equations 8.3 and 8.4.

$$NaOCl + H_2O \rightleftarrows Na^+ + OCl^- + H_2O \qquad (8.3)$$
$$Ca(OCl)_2 + H_2O \rightleftarrows Ca^{2+} + 2OCl^- + H_2O \qquad (8.4)$$

A portion of the available chlorine combines with ammonia and organic nitrogen compounds present in the water. This portion is called *combined available chlorine* and is represented by substances named *chloramines*. Their formation takes place according to the following equations:

$$NH_3 + HOCl \rightarrow NH_2Cl + H_2O \qquad (8.5)$$

$\uparrow$

Monochloramine

$$NH_2Cl + HOCl \rightarrow NHCl_2 + H_2O \qquad (8.6)$$

$\uparrow$

Dichloramine

$$NHCl_2 + HOCl \rightarrow NCl_3 + H_2O \qquad (8.7)$$

$\uparrow$

Trichloramine

The proportion of the three forms of chloramines greatly depends on the pH of the water.

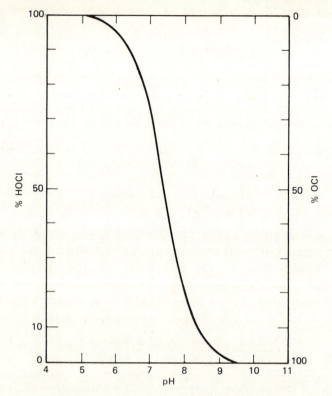

FIG. 8.1.   Distribution of HOCl⁻ and OCl⁻ in water as a function of pH. Adapted from G. M. Fair et al. (1948), *J. Am. Water Works Assoc.* **40**:1051.

- pH > 8.5: Monochloramine is predominant
- 4.5 < pH < 8.5: Coexistence of monochloramine and
dichloramine forms
- pH < 4.5: Trichloramine is the major form present

Since chloramines are less virucidal than hypochlorous acid, the presence of the latter in water is desirable for adequate disinfection. Therefore, one adds enough chlorine to oxidize the chloramine to nitrogen gas:

$$2NH_2Cl + HOCl \rightarrow N_2\uparrow + 3HCl + H_2O \qquad (8.8)$$

The level at which all the monochloramine is converted to nitrogen gas is called *breakpoint chlorination*. The addition of chlorine above this point will insure the existence of a free available residual.

Many factors are known to affect the chlorination process. These factors include chlorine species, concentration, temperature, pH, contact time, water turbidity, mixing, and the presence of interfering substances such as ammonia, organic nitrogen compounds, hydrogen sulfide, iron, or manganese.

Another chlorine species which has been evaluated for its disinfecting ability is chlorine dioxide (ClO_2).

ClO_2 is obtained according to equations 8.9 or 8.10.

$$2NaClO_2 + Cl_2 \rightarrow 2ClO_2 + 2NaCl \qquad (8.9)$$
$$2NaClO_2 + 4HCl \rightarrow 4ClO_2 + 5NaCl + 2H_2O \qquad (8.10)$$

Chlorine dioxide does not react with ammonia but it must be generated on site.

8.2.2 Iodine

Iodine is a good oxidizing halogen which hydrolyzes in water according to the equation:

$$I_2 + H_2O \rightarrow HOI + H^+ + I^- \qquad (8.11)$$

$\uparrow$ Iodine $\qquad \uparrow$ Hypoidous Acid $\qquad \uparrow$ Iodine

At pH 5–8, encountered in natural waters, iodine occurs mainly as I_2 or as HOI. The proportion of HOI and I_2 is controlled by the pH of the water. For example, at pH 5, 99% of iodine occurs as I_2 and 1% as HOI, whereas, at pH 8, 12% occurs as I_2 and 88% as HOI. Iodine does not interact with ammonia and nitrogenous organic compounds. Its action is less dependent than chlorine on contact time and temperature of the water. This halogen must, however, be used in higher concentrations than chlorine.

8.2.3 Bromine

For many reasons bromine has not become as popular as chlorine or even iodine for water disinfection. It is slightly soluble in water and it reacts with ammonia and nitrogenous organic compounds to form *bromamines*. Because of its physical properties this halogen is difficult to handle. A bromine species which is easier to handle and which has more oxidizing power is bromine chloride (BrCl). It quickly hydrolizes in water according to Equation 8.12.

$$BrCl + H_2O \rightarrow HOBr + HCl \qquad (8.12)$$

$\uparrow$ Hypobromous Acid

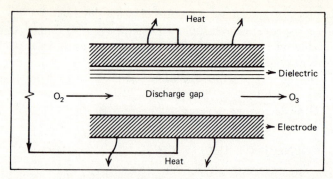

FIG. 8.2.   Basic ozonator configuration. From J. J. McCarthy and C. H. Smith (1974), *J. Am. Water Works Assoc.* **66**:718–725. Courtesy of American Water Works Association.

8.2.4   Ozone

Ozone (O_3) is a strong oxidizing agent which has been known for its germicidal properties since the nineteenth century. Ozone practice started in Europe, and its early use in water treatment plants was demonstrated in Paris, Nice, and Chartres, France. It was introduced in the United States to control tastes and odors. Because of its instability, ozone is generated on site by electrical discharge. An ozonator is made of electrodes separated by an air gap and a dielectric (Figure 8.2). Ozone is produced by passing dried air between the electrodes and by applying an alternating current with the voltage ranging from 8,000 to 20,000 V. The germicidal property of ozone is due to the liberation of free radicals (HO_2 and HO). Its effectiveness is not controlled by pH, and it does not interact with ammonia. It is generally used in the control of odors, taste, and color. It also helps in the precipitation of iron and manganese compounds. Advantages and disadvantages concerning its germicidal properties will be discussed later.

8.3   FATE OF VIRUSES FOLLOWING CHEMICAL DISINFECTION OF WATER

We have briefly reviewed the chemistry of oxidants used in the disinfection of water. We will now examine the effects of the various chemical species on viruses. It is generally agreed that proper disinfection of water will insure an adequate safeguard against microbial contamination. However, this is not always true under conditions prevailing in water and wastewater treatment plants. There are inherent differences among microbial pathogens with regard to their resistance to disinfectants. This resistance occurs in the following decreasing order:

> some bacterial spores > protozoan cysts > virus > bacteria

This raises questions concerning the adequacy of bacterial indicators for assessing the presence or absence of viruses in water. The search for a proper indicator is never-ending. This important topic will be examined in Chapter 12.

8.3.1   Chlorine

Of all the chlorine species which may form in water hypochlorous acid (HOCl) is known to be the most effective against viruses. Hypochlorite ion (OCl$^-$), which predominates at pH > 8 (see Figure 8.1), is not as efficient as HOCl. It is now well established that 0.5 to 1 ppm of hypochlorous acid will inactivate 99.99% of virus after a contact period of 30 minutes. There is, however, a wide variation in the resistance of enteric viruses to chlorine. Table 8.1 shows the resistance of 20 human enteric viruses to 0.5 ppm of free chlorine in the Potomac River. The time required to kill 99.99% of the virus varied from 2.7 minutes for reovirus type 1 to more than 60 minutes for echovirus 12. Chloramines, which are defined as combined residual chlorine, are much less efficient than free residual chlorine (about 50 times less efficient), and therefore they are needed in much higher concentrations to achieve an adequate disinfection of the water. The trichloramines are practically worthless as far as virus destruction is concerned. Hence whenever ammonia is present in the water one must go to breakpoint chlorination to insure proper inactivation of viruses.

Chlorine dioxide is a strong virucidal and bactericidal agent and has the advantage of not interacting with ammonia. Its virucidal effect increases as the pH is increased from 4.5 to 9.0. Chlorine dioxide at pH 7 is as efficient as hypochlorous acid at pH 6 with regard to poliovirus inactivation (Figure 8.3). It is, unfortunately, more expensive and must be generated on site.

Indigenous viruses may occur in clumps or may be embedded within organic and inorganic colloidal particles present in water and wastewater. The adsorption and aggregation processes afford some protection to viruses from the detrimental effect of chlorine or any other halogen. Recently, it was demonstrated that poliovirus occluded within fecal particulates was well protected from the effect of chlorine.

8.3.2   Iodine and Bromine

Iodine hydrolizes in water to give hypoiodous acid (HOI). Unlike hypochlorous acid, HOI predominates at relatively high pH (pH = 8) and displays some

TABLE 8.1.   Relative Resistance of 20 Human
Enteric Viruses to 0.5 mg/l Free Chlorine in
Potomac Water (pH 7.8 and 2°C)[a]

Comparison Based on First Order Reaction		Comparison Based on Experimental	
Virus	Min[b]	Virus	Min[b]
Reo 1	2.7	Reo 1	2.7
Reo 3	<4.0	Reo 3	<4.0
Reo 2	4.2	Reo 2	4.2
Adeno 3	4.8	Adeno 3	<4.3
Cox A9	6.8	Cox A9	6.8
Echo 7	7.1	Echo 7	7.1
Cox B1	8.5	Cox B1	8.5
Echo 9	12.4	Echo 9	12.4
Adeno 7	12.5	Adeno 7	12.5
Echo 11	13.4	Echo 11	13.4
Adeno 12	13.5	Polio 1	16.2
Echo 12	14.5	Echo 29	20.0
Polio 1	16.2	Adeno 12	23.5
Cox B3	16.2	Echo 1	26.1
Polio 3	16.7	Polio 3	30.0
Echo 29	20.0	Cox B3	35.0
Echo 1	26.1	Cox B5	39.5
Cox A5	33.5	Polio 2	40.0
Cox B5	39.5	Cox A5	53.5
Polio 2	40.0	Echo 12	>60.0

[a] Adapted from O. C. Liu et al. (1971), in: Virus and Water
Quality: Occurrence and Control. 13th Water Quality
Conf., Univ. of Illinois, Urbana–Champagne.
[b] Minutes required to kill 99.99% of virus.

virucidal property, although it is 4 to 5 times less efficient than HOCl. It has
the advantage of not reacting with nitrogenous compounds and this unique
property makes it a faster inactivating agent than chloramines. Probably the
most important application of iodine lies in the disinfection of swimming pools.
More than two decades ago it was also selected for the disinfection of canteen

water in the field and is known to be efficient against cysts of *Entamoeba histolytica*, which is responsible for amoebic dysentery in humans. However drinking water has a slight taste and color as a result of iodination.

The use of bromine in water disinfection has not been very popular because it is more expensive than chlorine. Little research has been carried out on its germicidal properties. Recent investigations have shown that bromine disinfection is controlled by the concentration of the chemical, contact time, temperature, and the extent of viral aggregation. Well dispersed poliovirus particles are rapidly inactivated with 10 μM of bromine in a matter of seconds (Figure 8.4). Other viruses have been shown to be rapidly inactivated by bromine.

8.3.3 Ozone

Ozone (O_3) is probably the best prospect in disinfection practice. Its quick virucidal effect does not depend on pH or ammonia content. Figure 8.5 shows the effect of two ozone concentrations on the survival of poliovirus type 1. At a level of 0.3 ppm, ozone is capable of reducing the virus titer by more than 99.9% in approximately 2 minutes.

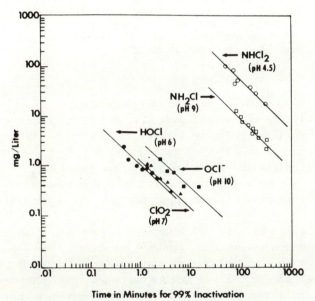

Time in Minutes for 99% Inactivation

FIG. 8.3. Comparison of the relative inactivation of poliovirus type 1 by hypochlorous acid, hypochlorite ion, monochloramine, dichloramine and chlorine dioxide at 15°C. From F. Brigano et al. (1979), in: *Progress in Wastewater Disinfection Technology*, A. D. Venosa, Ed., *Mun. Environ. Res. Lab.*, U.S. EPA, Cincinnati, Ohio.

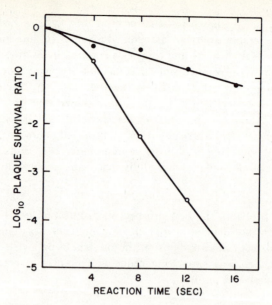

FIG. 8.4. Inactivation of poliovirus by bromine (temperature = 10° C); pH = 7). ●——● 1.9 μM; O——O 10 μM. Adapted from R. Floyd et al. (1976), *Appl. Environ. Microbiol.* **31**:298–303.

There is a *threshold ozone concentration* above which microbial inactivation is very rapid. For bacterial cells (*E. coli, B. cereus*) this critical concentration is around 0.1 mg/1 of ozone. Bacterial spores are more resistant to ozone and the threshold concentration is above 2 mg/1 (Figure 8.6). In viruses this critical concentration lies between 0.2 and 1 mg/1 of ozone (Figure 8.7).

Virus clumping can reduce the effectiveness of ozone treatment. Therefore sonication of water prior to ozone application generally enhances virus inactivation.

Unlike chlorine which reacts with the viral protein coat, it has been postulated that ozone destroys viruses by oxidizing the whole particle.

Proper disinfection of drinking water is undoubtedly the last line of defense against viruses. This goal may be reached only after adequate treatment of the water by chemical coagulation, sand, or diatomaceous earth filtration and adsorption to activated carbon. Pilot plant experiments have been undertaken in Laconia, N.H. to study virus removal during carbon adsorption followed by ozonation. This community draws its drinking water supply from Lake Winnipesaukee. It was shown that an ozone dosage of 1.13 mg/1 was sufficient to reduce coxsackievirus B3 by more than 99.999% (Table 8.2).

Since ozone does not leave any residual in water, it has been proposed to

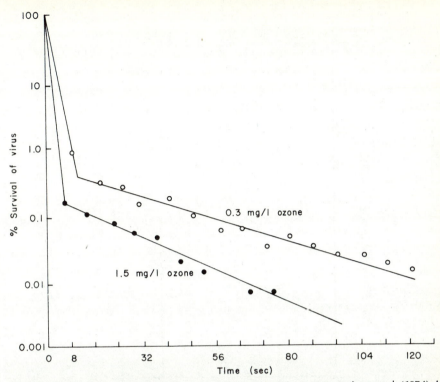

FIG. 8.5. Effect of ozone on poliovirus type 1. Adapted from E. Katzenelson et al. (1974), *J. Am. Water Works Assoc.* **66:**725–729.

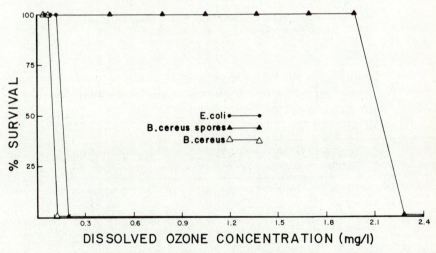

FIG. 8.6. Ozone threshold concentrations for bacteria (5-minutes contact time). Adapted from W. T. Broadwater et al. (1956), *Appl. Microbiol.* **26:**391–393.

185

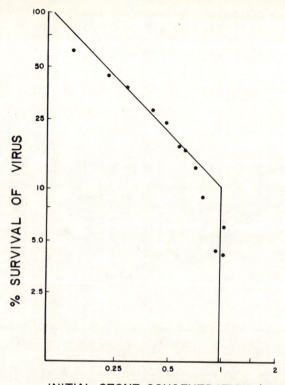

FIG. 8.7. Ozone threshold concentration for poliovirus (4-minutes contact time). Adapted from S. B. Majumdar et al. (1974), *J. Water Poll. Control Fed.* **46**:2048–2053.

TABLE 8.2. Pilot Plant Study of Coxsackie B3 Reduction Following Carbon Adsorption–Ozonation of Water from Lake Winnipesaukee[a]

Ozone Generated (ppm)	Ozone Dosage (ppm)	Total Virus Applied (PFU)	Virus Surviving (PFU)	Inactivation (%)
4.03	2.9	1.5×10^5	0	>99.99
2.0	1.53	7.6×10^7	0	>99.999
1.45	1.13	7.6×10^7	0	>99.999

[a] Adapted from J. W. Keller, R. A. Morin, and T. J. Schaffernoth (1974), *J. Am. Water Works Assoc.* **66**:730–733.

combine ozone treatment with postchlorination. This practice is carried out in the Netherlands.

8.4 FATE OF VIRUSES FOLLOWING CHEMICAL DISINFECTION OF WASTEWATER

8.4.1 Chlorine

Inactivation of Viruses by Chlorine in Wastewater

Chlorine has traditionally been used for the disinfection of wastewater effluents and continues to be popular despite its recently reported drawbacks.

Secondary wastewater effluents contain materials (suspended solids, organic materials, ammonia) that significantly reduce the efficiency of the chlorination process. This disadvantage can be overcome by pretreating the effluents by coagulation, flocculation, and filtration through sand or activated carbon. Furthermore, the elimination of ammonia via nitrification may eliminate the need for breakpoint chlorination. The presence of interferring substances in wastewater makes it imperative to use relatively high concentrations of chlorine (20–40 ppm) in order to achieve an adequate reduction of viruses. In wastewater effluents no free chlorine species are available after some seconds of contact. Thus the disinfecting ability of chlorine is mainly due to combined residuals such as monochloramine (NH_2Cl) or dichloramine ($NHCl_2$). The proportion of mono- and dichloramines is pH-dependent; monochloramines predominate at pH = 10 whereas dichloramines are the major species at pH = 6.0. Figure 8.8 illustrates the inactivation of poliovirus 3 at pH 6 and 10 by 10 and 30 mg/l of chlorine at 27°C. It appears from these data that dichloramines are more virucidal than monochloramines. A chlorine dosage lower than 10 mg/l is generally inefficient in the reduction of enteroviruses. Chlorination of primary or oxidation pond effluents does not result in a significant drop in virus numbers (Table 8.3). This is significant in view of the use of such effluents for agricultural purposes.

Virus monitoring in sewage treatment plants in Long Island has also revealed the presence of enteroviruses in chlorinated wastewater effluents. Virus concentrations varied from 67 to 4000 PFU/gal (Table 8.4).

Enteroviruses are generally more resistant to chlorination of wastewater than coliforms. As discussed previously, there are also differences in resistance to chlorine among enteroviruses Figure 8.9 shows that 2 and 20 mg/l of chlorine are necessary to achieve, in one hour, a 3 $\log_{10}$ reduction of coliforms and poliovirus, respectively.

Owing to their more frequent exposure to chlorine, natural virus isolates are thought to be more resistant to this chemical than laboratory strains. This

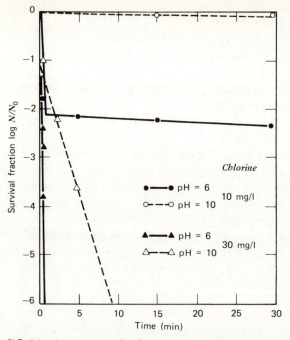

FIG. 8.8. Inactivation of poliovirus type 3 with chlorine in wastewater. (temperature = 27°C; no free chlorine was detected after 10 seconds. Combined chlorine residuals at 30 minutes after addition of 10 ppm chlorine were 2.1 and 2.5 ppm at pH 6.0 and 10.0, respectively. Combined chlorine residuals at 30 min. after addition of 30 ppm chlorine were 25.0 and 21.3 ppm at pH 6.0 and 10.0, respectively) Adapted from W. N. Cramer et al. (1976), *J. Water Poll. Control. Fed.* **48**:61–76.

increased resistance to chlorine has been demonstrated for a laboratory strain of poliovirus and is illustrated in Figure 8.10. The virus resistance increases after 10 cycles of exposure to 0.6–0.8 ppm of free residual chlorine.

Virus resistance to chlorination may also be due to clumping and to encapsulation of virus particles within particulate matter (cell debris and organic and inorganic colloidal matter). The survival of virus in aggregates may also be partly due to the phenomenon of *multiplicity reactivation.* This phenomenon consists of reactivation of stressed viruses (e.g., stress due to chlorine or UV radiation) through multiple infection of host cells. Laboratory experiments have revealed that some clays (e.g., bentonite) are also able to retard viral inactivation by chlorine. The protection of poliovirus type 1 from chlorine dioxide by bentonite appears to be temperature-dependent.

A study of this phenomenon under field conditions confirmed that solid-associated viruses are more resistant to chlorination than "free" virions. This indicates that one must take into account the solid-associated viruses when one develops a disinfection method. A reduction of turbidity to less than 0.1 JTU could be a preventive measure for counterbalancing the protective effect of particulate matter during disinfection.

It has been shown, at least for bacterial phage f2, that the RNA core is the main target of chlorine inactivation. The destructive action of a given chlorine species is apparently related to its ability to penetrate the protein coat.

Environmental Hazards Due to Chlorination of Wastewater

A survey of 400 sewage treatment plants in the United States has shown that chlorine is used for disinfection of wastewater in 74% of the plants. This is particularly significant, since many European countries rarely chlorinate their wastewater except when discharges are made into some particular areas such as shellfish beds or swimming areas. Thus in the United States serious concern

TABLE 8.3.   Fate of Viruses Following Chlorination of Primary and Oxidation Pond Effluents

Chlorine Level (mg/l)	Viruses (PFU/100 ml)	
	Prechlorination	Postchlorination
Primary Effluents[a]		
11.1	19.2	2.2
17.5	42.2	2.0
21.2	18.5	0.35
23.3	22.2	0.6
Oxidation Pond Effluents[b]		
8	45	43
8	100	100
8	45	30

[a] Adapted from G. Berg and T. G. Metcalf (1978), in: *Indicators of Virus in Water and Food*, G. Berg, Ed., Ann Arbor Sci. Pub., Ann Arbor, Michigan. (Temperature = 22–24°C; contact time = 15 minutes.)

[b] Adapted from Y. Kott (1975), in: *Discharge of Sewage from Sea Outfalls*, A. L. H. Gameson, Ed., Pergamon Press, Oxford, England. (Temperature = 20°C; contact time = 1 hour.)

TABLE 8.4. Virus Monitoring of Chlorinated Wastewater Effluents[a]

Sampling Date	Virus (PFU/gal)	Virus Isolated	Chlorine Residual (ppm)
A. Oyster Bay Sewage Treatment Plant[b]			
July 1976	227	Echo 11, 13, 14, 17, 25, 27 Coxsackie A7, A16, B3, B4, B6	—
Aug. 1976	No isolate		1.0
Sept. 1976	67.2	Polio 3 (vaccine strain) Coxsackie B2, B5 Echo 5, 6, 11, 12, 17, 25	—
Oct. 1976	No isolate		2.0
Mar. 1977	2636	Polio 2 (vaccine strain) Coxsackie B3 Echo 11	<0.2
April 1977	216	Coxsackie A17, B3 Echo 6	<0.2
B. Sunrise Sewage Treatment Plant[c]			
Aug. 1976	1440	Coxsackie B3, B4 Echo 6, 7, 21	2.0
Sept. 1976	1900	Polio 2 (vaccine strain) Coxsackie B2, B6 Echo 6, 7	2.0
Oct. 1976	854	Coxsackie A16 Echo 15, 31	2.0
Mar. 1977	990	Polio 1, 2 (vaccine strain) Coxsackie B4 Echo 6	<0.2
April 1977	4000	Polio 3 (vaccine) Coxsackie B3 Echo 2	<0.2

[a] Adapted from J. M. Vaughn and E. F. Landry (1977), "An Assessment of the Occurrence of Human Viruses in Long Island Aquatic Systems" Data Report, Brookhaven Nat. Lab., Upton, N. Y.

[b] Trickling filter effluent.

[c] Contact stabilization effluent.

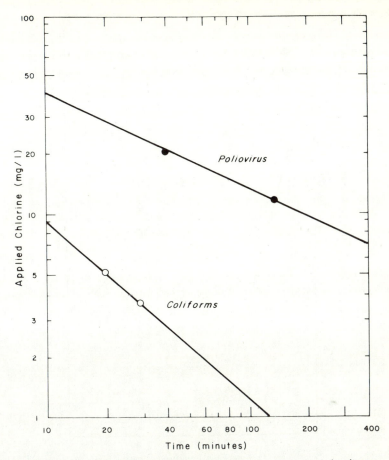

FIG. 8.9.   Concentration time relationship for 99.9% inactivation of poliovirus and coliforms in sewage effluent by chlorine at 20°C. Adapted from H. Shuval (1974), in: *7th Int. Conf. Water Poll. Res., Paris.*

has been raised over the harmful effects of chlorinated discharges. These effects fall into two categories:

1. *Toxicity to Aquatic Life.* This is translated into a reduction of primary producers and into fish kills or impaired reproduction and growth.

2. *Toxicity to Humans.* The discharge of chlorinated effluents may result in the formation of chlorinated hydrocarbons (e.g., chloroform, carbon tetrachloride), some of which are carcinogenic to laboratory animals and possibly to man. A study of New Orleans' drinking water supply

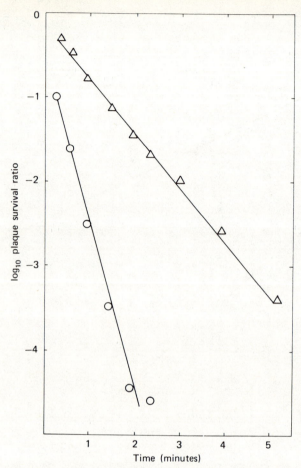

FIG. 8.10.   Rate of inactivation of poliovirus during cyclic exposure to chlorine. O——O cycle one, Δ——Δ cycle 10. Adapted from R. C. Bates et al. (1977), *Appl. Environ. Microbiol.* **34**:849–853.

revealed the presence of many of these chlorinated hydrocarbons. This problem may be solved by:

- *Removing chlorine before discharge.* Dechlorination can be achieved with sulfur dioxide or activated carbon treatment, but this adds a 30% extra cost.

- *Reducing sewage chlorination.* Recommended by a 1977 report to the U.S. Congress.

- *Finding alternatives to sewage chlorination.*

8.4.2 Iodine

We have seen that, about 2 decades ago, iodine was proposed for the disinfection of swimming pools. More recently treatment of wastewater effluents (activated sludge effluents) with iodine has been studied. There is evidence that iodine, at a concentration of 30 mg/l, is more virucidal at high than at low pH (Figure 8.11). Contrary to chlorine action, nucleic acid is not the target of iodine inactivation. Iodine rather reacts with the viral protein coat. For example, the destruction of bacterial phage f2 with this chemical is due to iodination of tyrosine in the protein coat.

8.4.3 Ozone

It was previously shown that ozone has been used in Europe, particularly in France, since the turn of the century for the disinfection of drinking water. However, it was only until a decade ago that this powerful oxidant was first

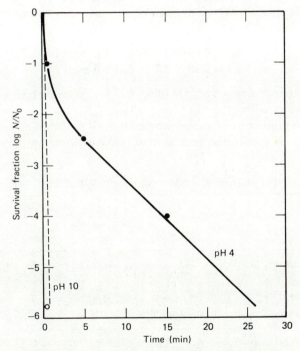

FIG. 8.11. Effect of iodine on poliovirus 3 inactivation in wastewater (iodine conc. = 30 ppm; temperature = 27°C) Adapted from W. N. Cramer et al. (1976), *J. Water Poll. Contr. Fed.* **48**:61–76.

used in England for the disinfection of domestic wastewater. Ozonation became then popular in the United States, and wastewater ozonation units are already in operation in some cities such as Chicago, Ill., Blue Plains, Washington, D.C., and Indiantown, Fla. Many other facilities are under construction or in the design stage.

Although wastewater ozonation is known to substantially reduce color, cyanides, and other chemicals, comparatively little is known about virus inactivation. In relatively clean water ozone, at a concentration of less than 1 ppm, is capable of achieving a 3 $\log_{10}$ reduction of viruses in a matter of seconds. However, in wastewater ozone reacts with organic materials, and higher chemical concentrations and contact time are thus required to achieve an adequate disinfection. Table 8.5 shows that 4 to 15 ppm of ozone are necessary to bring about 99.99 to 100% reduction of viruses within a period of 1 to 5 minutes.

As discussed for drinking water, ozone exerts an *all-or-none effect*; this means that there exists a *threshold concentration* of ozone above which virus inactivation is very rapid. This threshold concentration is around 1 ppm. The relationship of virus survival with ozone concentration and contact time is given by the following:

$$Ct = 0.01 \ S^{-1.12} \text{ at } C < 1.0 \text{ mg/l} \qquad (8.13)$$
$$Ct = 0.13 \ S^{-0.36} \text{ at } C > 1.0 \text{ mg/l} \qquad (8.14)$$

where C = ozone concentration, t = time, and S = surviving fraction of viruses. This relationship is illustrated in Figure 8.12.

It has been proposed to combine sonication with ozonation to achieve higher virus inactivation. Sonic energy is able to break up viral aggregates and thus

TABLE 8.5.   Virus Inactivation Following Wastewater Ozonation[a]

Wastewater	Virus	Ozone Concentration (ppm)	Contact Time (min)	% Inactivation
Package treatment plant effluent	f2	15	5	≃100
Primary effluent	Poliovirus 2	4.44	1	99.99
Secondary effluent	Poliovirus 2	5.05	1	99.99

[a] Adapted from J. L. Pavoni and M. E. Titlebaum (1974), *in: Survival of Viruses in Water and Wastewater Systems*, J. F. Malina, Jr. and B. P. Sagik, Eds., University of Texas at Austin, and from S. B. Majumdar, W. H. Ceckler, and O. J. Sproul (1974), *J. Water Poll. Control Fed.* **46**:2048–2053.

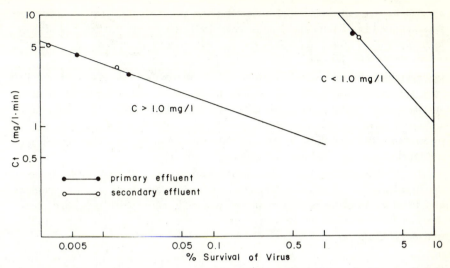

FIG. 8.12. Effect of contact time and ozone concentration on poliovirus survival in wastewater effluents. From S. B. Majumdar et al. (1974), *J. Water Poll. Contr. Fed.* **46**:2048–2053. Courtesy of Water Pollution Control Federation.

expose single virions to the action of ozone. A treatment plant utilizing this combination is in operation in Indiantown, Florida.

With regard to the site of ozone inactivation it has been suggested that this chemical acts on viruses by oxidizing the whole virus particle.

Although ozone is expensive and leaves no residual, it is regarded as the best alternative to chlorine, since it is a good virucidal agent that is not known to be toxic to the aquatic biota.

8.5 PHYSICAL DESTRUCTION OF VIRUSES

8.5.1 Ultraviolet Radiation

Short wave ultraviolet radiation (2000–2950 Å) effectively destroys microorganisms, including viruses. We have examined, in Chapter 3, the mode of action of UV on microorganisms. Since the short-wave UV radiation is not naturally produced at the earth's surface, it is usually generated in low pressure mercury vapor lamps where electrical energy is converted to UV. These lamps are made of quartz to allow passage of UV radiation at the germicidal wavelength of 2537 Å.

The use of ultraviolet radiation in the disinfection of water is almost as old as chlorination. In the United States the first water treatment plant using UV was installed in Henderson, Kentucky in 1916. However, this method of disinfection was abandoned in favor of chlorine which was much less expensive.

Water disinfection with UV radiation is generally efficient with regard to bacteria and viruses. Total inactivation of enteroviruses is generally reached with relatively clean water and at flow rates not exceeding the manufacturer's specifications (Table 8.6). The effectiveness of UV radiation is reduced by turbidity and color. Therefore, water must be filtered to remove the interference due to solids. There are now many ultraviolet water purifiers on the market and a typical one is illustrated in Figure 8.13.

Despite the fact that UV leaves no residual in water, it does not have any of the disadvantages of chlorine. It does not produce any taste or color problem, and no toxic residual is formed in the water. Further work is needed on the effect of UV radiation on microorganisms in wastewater and on the cost-effectiveness of the process.

8.5.2 Photodynamic Inactivation

Microorganisms, including viruses, are rendered extremely sensitive to visible light when in the presence of photosensitizing chemicals such as dyes

TABLE 8.6. Effect of Ultraviolet Radiation on Viruses[a]

Type of Water	Virus Type (Initial Titer)	Flow Rate (liters/min)	Extent of Virus Inactivation
Seine River water	Poliovirus 1 (10^3 $TCID_{50}$ ml)	8.3	Total inactivation
Springwater	Poliovirus 1 (10^3 $TCID_{50}$ ml)	13.3	Total inactivation
Springwater	Poliovirus 3	13.3	Total inactivation
Seine River water	Poliovirus 3	13.3	4 log reduction
Demineralized water	Echovirus 7 (1.2×10^2 PFU/ml)	34–58	Total inactivation
Demineralized water	Coxsackie A9	34	99.2%
Demineralized water	Poliovirus 2 (1.5×10^3 PFU/ml)	34	Total inactivation
Demineralized water + 9 color units	Poliovirus 2 (1.3×10^3 PFU/ml)	34	98.4%

[a] Adapted from J. Maurin, L. P. Mazoit, A. Dodin, and G. Escallier (1974), *Bull. WHO* **51**:35–39; and C. B. Huff, H. F. Smith, W. D. Boring, and N. A. Clarke (1965), *Publ. Health Rep.* **80**:695–705.

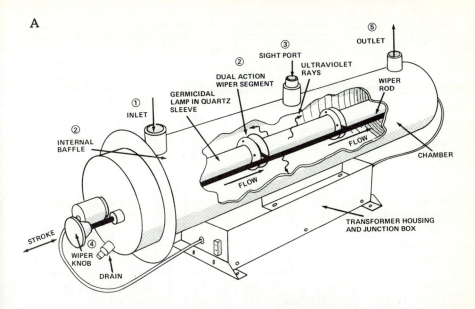

FIG. 8.13.   (A) Diagram of an ultraviolet water purifier. (B) The Sanitron ultraviolet water purifier. Courtesy of Atlantic Ultraviolet Corp. Bay Shore, N.Y.

(methylene blue, neutral red). This phenomenon, termed *photodynamic inactivation,* has been proposed recently for the destruction of bacterial and viral pathogens in sewage. The outcome of photodynamic action on microorganisms depends on dye concentration, pH, temperature, sensitization time in the dark, and the turbidity of the sewage. Nearly 6 $\log_{10}$ reduction in poliovirus titer is

TABLE 8.7.   Photodynamic Inactivation of Poliovirus 1[a]

		Poliovirus Titer, log_{10} PFU/ml			
	Sensitization Time	Unfiltered Sewage Effluent		Sand-filtered Effluent	
Treatment	(hr)	Light	Dark	Light	Dark
Sewage	0	6.2	6.0	5.8	5.9
effluent pH	16	5.9	6.1	4.4	4.5
= 10.0					
Sewage	0	6.0	6.2	5.8	6.1
effluent pH	16	3.9	5.9	0.0	4.4
10 + 5 mg/l					
methylene					
blue[b]					

[a] Adapted from C. Wallis, C. P. Gerba, and J. L. Melnick (1976), in: Viruses in Water, G. Berg et al., Eds., APHA, Washington, D.C.

[b] Samples were held at 25°C for the time indicated and then exposed to 8000 μW/cm² for 60 seconds.

achieved at 25°C in sand-filtered sewage, pH 10, and treated with 5 mg/l methylene blue for a period of 16 hours in the dark (Table 8.7).

Although this process appears efficient in viral destruction, its public health implications have not been properly evaluated. Moreover, although methylene blue can be removed from sewage effluents by activated carbon, it is not known whether traces of this dye reaching our natural waters can lead to health hazards to the aquatic biota and eventually to man.

8.6   SUMMARY

1. Chlorine is the major disinfectant used in water and wastewater treatment plants.

2. Among all the chlorine species formed in water, hypochlorous acid (HOCl) is the most effective against viruses. In wastewater the disinfecting ability of chlorine is due mainly to combined residuals such as chloramines. Dichloramines are more virucidal than monochloramines. The monitoring of sewage treatment plants for viruses has revealed the presence of viruses in chlorinated effluents.

3.   Arguments have been raised against chlorination of wastewater effluents. Chlorinated discharges are implicated in toxicity to aquatic life and, perhaps, in causing cancer in humans.

4.   Ozone is probably the best available alternative to chlorination of water. It is a strong virucidal agent but, unfortunately, leaves no residual in the water. It has been proposed to combine ozone treatment with postchlorination.

5.   Ultraviolet irradiation has been proposed for the disinfection of water and wastewater. It is generally efficient against viruses but the water must, however, be pretreated for removal of turbidity and color.

6.   There is an urgent need to conduct disinfection studies on infectious hepatitis A virus and gastroenteritis virus(es).

FURTHER READING

American Water Works Association. 1971. *Water Quality and Treatment.* McGraw-Hill, New York.

Berg, G., H. L. Bodily, E. H. Lennette, J. L. Melnick, and T. G. Metcalf, Eds. 1976. *Viruses in Water.* American Public Health Association, Washington, D.C.

Hoehn, R. C. 1976. Comparative disinfection methods. *J. Am. Water Works Assoc.* **68:**302–308.

McCarthy, J. J., and C. H. Smith. 1974. A review of ozone and its applications to domestic wastewater treatment. *J. Am. Water Works Assoc.* **66:**718–725.

Malina, J. F., Jr., and B. P. Sagik, Eds. 1976. *Virus Survival in Water and Wastewater Systems.* University of Texas Press, Austin.

Snoeyink, V. L., et al., Eds. 1971. *Virus and Water Quality: Occurrence and Control.* 13th Water Quality Conf., University of Illinois, Urbana–Champaign.

nine
Fate of Viruses in Land Disposal of Wastewater Effluents

9.1   INTRODUCTION

Sewage application onto land is hardly a novelty. For centuries, the Chinese have been applying untreated human wastes or *night soil* onto land. This practice was considered a good method for supplying soils with valuable nutrients. In England, in the nineteenth century, the first Royal Commission on Sewage Disposal stated that "the right way to dispose of town sewage is to apply it continuously to land, and it is why, by such applications, that the pollution of rivers can be avoided." In the United States Massachusetts was the first state to try irrigation of land with sewage (1870s). Until the 1940s, only the nutritional aspect of sewage application to land was considered, and the health aspects were generally of no concern. Since then we know that sewage carries many agents that are pathogenic to both humans and animals. The most important pathogens and parasites include enteric viruses, bacteria (*Salmonella, Shigella, Mycobacterium, Leptospira*) parasitic protozoa, and helminths.

The pathogen load of a given sewage varies from one community to the other and depends on the population density and sanitary habits of the people in the community. In the United States the pathogenic load of sewage is generally low as compared with other countries where water shortages are frequent and the sewage is more concentrated. This has led some to suggest that the potential for the spread of infection as a result of sewage application onto land is low. However, we do not have any solid epidemiological data to confirm or negate these statements. Examples taken from around the globe have shown that a heavily contaminated working environment may be conducive to disease among humans and animals.

If one contemplates the application of sewage effluents onto land one must reduce the pathogenic load as much as possible by proper treatment and disinfection. Table 9.1 gives an estimation of pathogen load applied to soil following secondary treatment and disinfection.

In recent years increasing concern over the quality of surface waters has led to a search for other sewage disposal methods. It now appears that land application of sewage effluent is a good viable alternative. The Water Pollution Control Act Amendment of 1972 (PL 92-500) states that "Waste treatment management plans and practices should provide for the application of the best

TABLE 9.1. Estimation of Wastewater Pathogens Applied to Soil[a,b]

Pathogen	Total Number in Raw Wastewater	Total Number after Treatment[c]	Organisms (applied per acre per day)
Salmonella	2×10^{10}	5×10^5	3.9×10^3
Mycobacterium	2×10^8	1.5×10^4	1.2×10^2
E. histolytica	1.5×10^7	1.2×10^4	9.3×10^1
Helminth ova	2.5×10^8	5×10^3	3.9×10^1
Virus	4×10^{10}	2×10^6	1.6×10^4

[a] Adapted from D. H. Foster and R. S. Engelbrecht (1973), in: Recycling Treated Municipal Wastewater and Sludge Through Forest and Cropland, W. E. Sopper and L. T. Kardos, Eds., Pennsylvania State University Press, University Park, Pa.

[b] Wastewater applied at a rate of 2 inches per week.

[c] Total number after sewage treatment which consists of primary and secondary treatments followed by disinfection.

practicable waste treatment technology before any discharge into receiving water. . . ." It also states that ". . . the groundwater that would result from wastewater application should not exceed bacteriological and chemical levels for raw or untreated drinking water supplies."

Land application of sewage effluents and residuals has many benefits, the most important being:

- Water conservation in the form of aquifer recharge.
- Protection of surface waters from fecal, organic, and other types of pollution.
- Enrichment of soil with organic matter.
- Supply of valuable inorganic nutrients (N, P, K) to crops.

However, potential problems are associated with this practice. This chapter deals with the examination of these problems, especially those of viral origin.

9.2 METHODS OF LAND APPLICATION OF SEWAGE EFFLUENTS

Three basic methods (Figure 9.1) are used in the application of sewage effluents to land. The choice of a given method depends on the conditions pre-

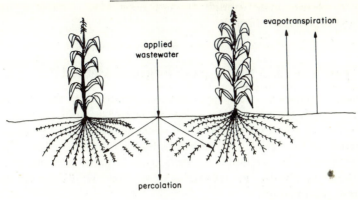

A. Low Rate Land Treatment

applied
wastewater

evapotranspiration

percolation

B. Overland Flow

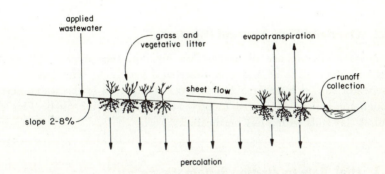

applied
wastewater

grass and
vegetative litter

evapotranspiration

runoff
collection

sheet flow

slope 2-8%

percolation

C. Rapid Infiltration

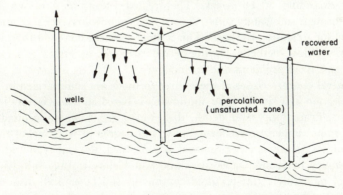

recovered
water

wells

percolation
(unsaturated zone)

FIG. 9.1.   Methods of land application of sewage effluents. From Manual for Land Treatment of Municipal Wastewater. U.S. Environ. Protect. Agency (1977).

vailing at the site under consideration (loading rates, method of irrigation, crops, and expected treatment).

9.2.1   Low Rate Land Treatment or Irrigation Method (Figure 9.1A)

The main objectives of this method are to renovate water and produce crops. Sewage effluents are applied via sprinkling or surface application at a rate of 1.5 to 10 cm/week. Two thirds of the water is taken up by crops or lost by evaporation and the remainder percolates through the soil matrix. The system must be designed to maximize denitrification in order to avoid pollution of groundwater by nitrates. Phosphorus is immobilized within the soil matrix by fixation or precipitation.

The irrigation method is mainly used by small communities and requires large areas, generally in the order of 5–6 hectares per 1000 people.

9.2.2   Overland Flow Method (Figure 9.1B)

Wastewater effluents are allowed to flow for a distance of 50 to 100 m along a vegetated slope (2–8%) and are collected in a ditch. The loading rate of wastewater ranges from 5 to 14 cm/week. Only 10% of the water percolates through the soil as compared to 60% running off to the ditch. This system requires clay soils with a low permeability.

9.2.3   High Rate Infiltration System (Figure 9.1C)

The primary objective of this method is the treatment of wastewater at loading rates often exceeding 50 cm/week. The treated water, most of which has percolated through soil (sandy soils), is used for groundwater recharge and may be recovered for irrigation purpose. This system requires less land than irrigation or overland flow methods. Drying periods are often necessary to aerate the soil system and avoid problems due to clogging.

Some general characteristics of these three methods are summarized in Table 9.2. The selection of a site for land application is based on many factors including soil type, drainability and depth, distance to groundwater, groundwater movement, slope, underground formations, and degree of isolation of the site from the public. As far as virus removal is concerned, the soil type and depth and the distance to groundwater are important factors to take into account. Some feet of moderately fine-textured continuous soil layer and an adequate distance to groundwater are necessary for virus removal.

TABLE 9.2. General Characteristics of the Three Methods Used for Land Application of Sewage Effluents[a]

Factor	Irrigation Method	Overland Flow Method	High Rate Infiltration Method
Main objectives	Reuse of nutrients and water, Wastewater treatment	Wastewater treatment	Wastewater treatment, Groundwater recharge
Soil permeability	Moderate (Sandy to clay soils)	Slow (Clay soils)	Rapid (Sandy soils)
Need for vegetation	Required	Required	Optional
Loading rate	1.3–10 cm/wk	5–14 cm/wk	>50 cm
Application technique	Spray, surface	Usually spray	Usually spray
Land required for 1 MGD flow (⊦ buffer zone)	62–560 acres	46–140 acres	2–62 acres
Needed depth to groundwater	About 5 ft	Undetermined	About 15 ft
BOD and suspended solid removal	90–99%	90–99%	90–99%
N removal	85–90%	70–90%	0–80%
P removal	80–90%	50–60%	75–90%
Use to recharge groundwater	0–30%	0–10%	up to 90%

[a] Adapted from R. L. Wright (1975), *Florida Sci.* **38**:207–222, and from C. E. Pound et al. (1978), *Civil Eng.* June 1978.

9.3 EXAMINATION OF HEALTH PROBLEMS ASSOCIATED WITH LAND APPLICATION OF SEWAGE EFFLUENTS

Concern has been raised over the contamination of groundwater with pollutants of chemical and microbial origin during land application of wastewater effluents.

9.3.1   Contamination of Groundwater with Excess Nitrate

Phosphorus in sewage effluent is of no major concern because it is easily removed from water by crops, adsorption to soil and/or by precipitation after reacting with Al, Fe, or Ca.

Most of the nitrogen present in secondary waste effluents is the form of ammonium (NH_4^+). The ammonium nitrogen undergoes the process of *nitrification* which results in the production of nitrate (NO_3^-). This process is catalyzed by specialized autotrophic microorganisms belonging to the genera *Nitrosomonas* and *Nitrobacter*:

$$NH_4^+ \xrightarrow{\text{Nitrosomonas}} NO_2^- \xrightarrow{\text{Nitrobacter}} NO_3^-$$

Ammonium   Nitrite   Nitrate

Nitrification operates under aerobic conditions. Nitrate can then be assimilated, denitrified (production of N_2) or it may travel through the soil and reach the groundwater supplies. The transport of nitrates through soils is particularly important in sandy soils. Excess nitrate in groundwater used for drinking purposes may be responsible for a disease called *Methemoglobinemia,* which strikes infants under 4 months of age. Nitrate is reduced to nitrite by bacteria. Nitrite competes with O_2 for hemoglobin and forms *methemoglobin,* which is unable to carry oxygen and, thus, leads to asphyxia. The U.S. Public Health standard for nitrate in drinking water has been set at 45 ppm as NO_3 .

9.3.2   Contamination of Soils and Groundwater with Heavy Metals

Heavy metals (Cd, Hg, Pb, Ni, Cr) are not known to be of any use in plant and animal metabolism. It is likely that these metals concentrate in plant materials and then pass into the food chain. Their mobility and solubility depend on the physicochemical conditions within the soil matrix. For example, an increase in pH decreases their solubility and hence their transport through soil. This is the reason why liming is recommended for that purpose. Domestic sewage effluents contain generally low amounts of heavy metals. The main concern is over their concentration in sludge and this topic will be covered in Chapter 10 in more detail.

9.3.3   Contamination by Microbial Pathogens

There are three basic problems associated with land application of sewage effluents:

1. *Microbial aerosols generated by spray irrigation.*
2. *Contamination of crops.* This problem is particularly acute in the case of vegetables which are eaten raw.
3. *Contamination of groundwater and surface waters.*

Pathogens (bacteria, viruses, protozoan cysts and parasitic worms) will have to travel through the soil matrix in order to reach the groundwater supply. This is the subject of the next section.

9.4   FATE OF VIRUSES IN SOILS

9.4.1   Introduction to Soil Science

Before discussing virus retention by soils, it seems appropriate to give the most relevant information concerning soil properties which mostly control the survival and transport pattern of viruses. Through land application of sewage effluents and sludges, viruses come into contact with the soil matrix and their persistence and/or movement will be readily controlled by some properties of that complex environment.

Soils are very dynamic and complex systems formed through the influence of physical, chemical and biological factors. Five main factors contribute to soil formation and these are: climate (temperature, rainfall), biota (plants, animals, microorganisms), topography, parent material and finally, time. Soils (Figure 9.2) are made of about 50% solids and 50% pore space. Approximately half of

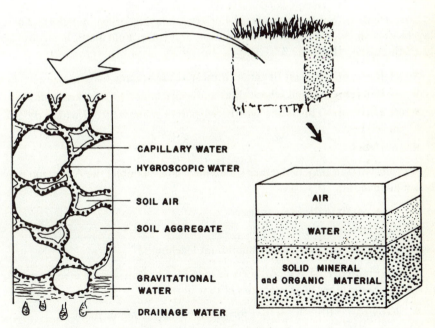

FIG. 9.2. Physical composition of soils. From K. C. Berger (1965), *Introductory Soils*, Macmillan Co., New York. Courtesy of MacMillan Co.

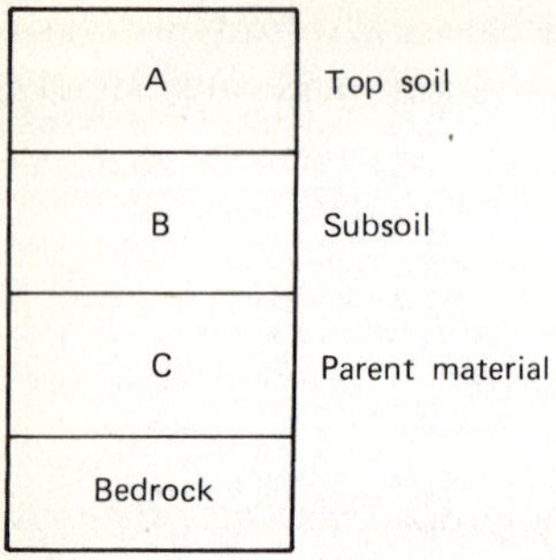

A	Top soil
B	Subsoil
C	Parent material
Bedrock	

FIG. 9.3.   A soil profile.

the pore space is occupied by water and the other half by air. The solid phase is made of organic ($<1\%$ to more than 95%) and mineral matter (gravels, sand, silt, and clay particles).

Some Definitions

This information is useful when one conducts laboratory experiments on virus transport through soil columns. It is also necessary for the proper understanding of virus migration through soils under field conditions.

Soil Profile. Soils are made of horizontal layers called *horizons*. These layers vary in thickness, texture, structure, and color. Four main horizons can be distinguished (Figure 9.3).

- *A Horizon.* Topsoil (it contains most of the organic matter).
- *B Horizon.* Subsoil where clays and other materials may accumulate.
- *C Horizon.* This is the weathered parent material from which the soil developed.
- *Bedrock*

A given horizon may be further subdivided. For example, the A horizon may be subdivided into

- A_0. Containing organic matter.
- A_1. Dark colored.
- A_2. Zone characterized by maximum leaching.

The capital letter for a given horizon may also be followed by a suffix which designates some other characteristics of that horizon.

- *Example*:   Bh: "h" stands for accumulation of humic material.
               Bt: "t" stands for the accumulation of clay.

Texture. It is a soil characteristic which indicates the proportion of components of the mineral fraction. These components are coarse sand, fine sand, very fine sand, silt and clays. Their sizes are shown in Table 9.3. Once we measure the proportions of the different components, one refers to a *soil textural triangle* and reads the texture of the soil (Figure 9.4). Clay minerals, because of their colloidal size, are the most active component of the mineral fraction. They will be later discussed separately.

Structure. The structure designates the amount of aggregation between soil particles.

Particle Density. Particle density is expressed in grams/cc and it is assumed that it is approximately equal to 2.65.

Bulk Density. Bulk density is the weight of a given volume of soil. It is also expressed in g/cc.

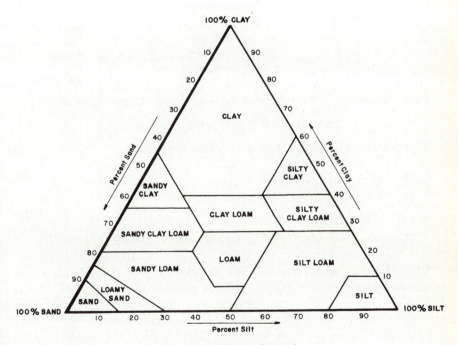

FIG. 9.4. Soil textural triangle.

TABLE 9.3.   Size of Individual Components of the Soil Mineral Fraction

Soil Component	Size (mm)
Very coarse sand	2.00–1.00
Coarse sand	1.00–0.50
Medium sand	0.50–0.25
Fine sand	0.25–0.10
Very fine sand	0.10–0.05
Silt	0.05–0.002
Clay	<0.002

% Pore Space. Pore space is the percent of a given volume of soil not occupied by solids.

$$\% \text{ solid space} = \frac{\text{bulk density}}{\text{particle density}} \times 100$$

$$\% \text{ pore space} = 100 - \frac{\text{bulk density}}{\text{particle density}} \times 100$$

Pore Volume. It is the volume of the water inside the pore space of the soil.

$$\text{Pore volume} = \% \text{ pore space} \times \text{soil volume}$$

This parameter is often used in soil leaching experiments.

Colloidal Properties of Soil

There are two types of colloids in soils systems:

* *Mineral Colloids.* Clay minerals.
* *Organic Colloids.* Humic materials commonly termed "humus."

Clay Minerals and Their Surface Properties. Clay minerals are aluminosilicates of colloidal size and well defined crystallographic structure. Due to their high specific surface area, they exert a direct influence on the physicochemical properties of the soil and on the biological processes as well.

Clay minerals are made of structural layers of SiO_2 and Al_2O_3–$Al(OH)_3$ (Figure 9.5). The proportion between these 2 types of layers serves as a criterion for the classification of clay minerals. Thus, one distinguishes between the two-layer types and the three-layer types of clay minerals. In the former the ratio SiO_2 layer : Al_2O_3-$Al(OH)_3$ layer is 1, whereas in the latter this ratio is 2 (Figure 9.5; Table 9.4).

Within the clay lattice, there are always substitutions of Si and Al by other cations of similar size but lower valency. The replacement of Si by Al in the silica layers, and of Al by Fe, Mg, Ni, and so on, in the alumina layers, is known as *isomorphic substitution*. A direct consequence of this substitution is a

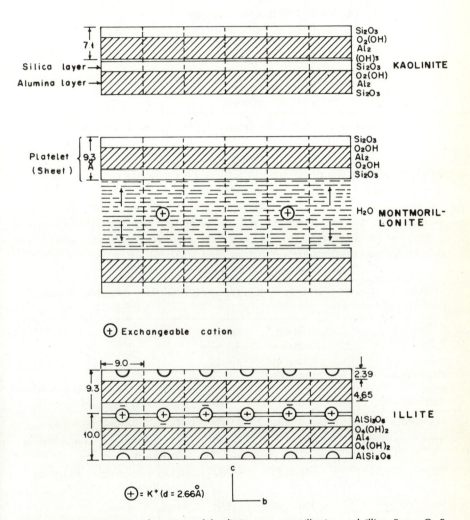

FIG. 9.5. Diagrammatic edge view of kaolinite, montmorillonite and illite. From C. E. Marshall (1964), *Physical Chemistry and Mineralogy of Soils*, Vol. 1, Wiley, New York. Courtesy of Wiley.

TABLE 9.4.   Classification of Clay Minerals[a]

A. *Amorphous*
   1. Allophane group
B. *Crystalline*
   1. Two-layer types
      a. Kaolinite, Halloysite
   2. Three-layer types
      a. Expanding lattice: montmorillonite,
         vermiculite, nontronite, saponite, hectorite
      b. Nonexpanding lattice: illite group
   3. Regular mixed-layer types
      a. Chlorite group
   4. Chain-structure types
      a. Attapulgite, sepiolite

---

[a] Adapted from R. E. Grim (1968), *Clay Mineralogy*, McGraw-Hill Co., New York.

deficit of positive charge of the clay lattice. This deficit is compensated by the absorption of an equivalent amount of cations (e.g., Ca, Mg, K, Na, etc.) on the clay lattice. The ability of clays to adsorb ions and retain them in an exchangeable position is termed *ion exchange* and the ions involved in this process are *exchangeable cations* (Table 9.5). The exchange sites resulting from isomorphic substitutions are located on the planar surface of the clay minerals.

TABLE 9.5.   Cation Exchange Capacity, Specific Surface Area and Shape of Some Clay Minerals[a]

Clay Type	Cation Exchange Capacity (meq/100 g)	Specific Surface Area (m²/g)	Shape
Kaolinite	3–15	48	Hexagonal particles
Halloysite	5–10	75	Elongate tubular particles
Montmorillonite	80–150	810	Particles with no characteristic outlines
Illite	10–40	192	Flakes with crude hexagonal outlines
Vermiculite	100–150	550	Similar to illite
Attapulgite	3–15	—	Elongate lath-shaped units

[a] Adapted from R. E. Grim (1968), *Clay Mineralogy*, McGraw-Hill Co., New York.

Owing to broken bonds, the crystal edges have also the ability to exchange ions with the external solution. However, whereas the electric charge resulting from isomorphic substitution is independent of the pH, the charge at the edges of clay particles is pH-dependent, being positive at low pH values and negative at higher pH.

So far the properties of one clay platelet have been described. Separate platelets can be obtained in montmorillonite under certain conditions. Clay platelets are well dispersed in Na-montmorillonite suspensions, whereas in Ca-montmorillonite they form *tactoids* consisting of 5 to 15 parallel plates.

Progress in electron microscopy has permitted a better appreciation of the shape and size of clay minerals (Table 9.5). Electron micrographs generally show well-formed six-sided flakes for kaolinite, elongate tubular particles for halloysite, extremely small particles with no characteristic outlines for mont-morillonite and small flakes with crude hexagonal outlines for illite. Particle-size studies show extremely small particles for montmorillonite (0.2 μm), 0.1–0.3 μm for illite, and 0.3–4 μm for kaolinite.

The different clay minerals show marked differences in their specific surface area (Table 9.5). The latter varies from 50 m²/g of clay for the two-layer minerals, to 800 m²/g for the three-layer ones.

In many of the three-layer minerals water can enter between the unit layers, giving these minerals an expanded lattice structure, e.g., montmorillonite may expand along the c axis from 9.6 to 21.4 Å. In the presence of K^+ ions, the lattice may collapse along the *c* axis trapping these ions between the unit layers, e.g., illite which is a three-layer nonexpanding clay. In contrast, the two-layer minerals do not expand and do not hold interlayer water or interlayer cations, e.g., kaolinite (Figure 9.5).

Organic Colloids. Organic matter in soils is the product of decomposition of plant and animal materials by microorganisms. It may represent from less than 1% to more than 95% of the soil weight. Well-decomposed organic matter is called *humus* and is colloidal in size. The exact chemical composition of humus is not known and it is generally composed of liguin, proteins, and polysaccharides (polyuronides).

Organic matter has many functions in soils:

- It is involved in cation exchange reactions. The cation exchange capacity of humic material reaches 200 meq/100 g soil.
- It affects the physical properties of soils such as structure, texture, water-holding capacity, and aeration.

However, the role of organic matter is not completely understood, particu-larly its effect on the survival and transport pattern of bacteria and viruses in soils.

9.4.2 Retention of Viruses by the Soil Matrix

The brief introduction to soil science has permitted us to better understand the different components of soils and their relationships. We now know that soil is a complex environment where the solid phase plays a major role in various processes occurring within the soil matrix.

Virus Transport Through Soils Under Field Conditions

Field studies are essential in the determination of the fate of viruses in land treatment operations. Their undertaking was made possible only after the development of adequate methods for detecting low numbers of viruses in large volumes of water (see Chapter 5). It is now possible to process on site more than a 100 gallons of groundwater pumped from monitoring wells within the land disposal site. Field studies are an expensive undertaking and they are sometimes plagued with problems due to weather conditions, filter clogging, sample transportation, and storage.

One of the earliest field studies involving a virus monitoring program is the Santee Project near San Diego, California. Sewage effluents were pumped into percolation beds made of sand and gravel. No virus was found in monitoring wells located 200 feet from the infiltration site.

The Flushing Meadow project near Phoenix, Arizona involved the renovation of activated sludge effluents by high-rate infiltration (about 100 m/year) through a fine loamy sand underlain by sand and gravel to a depth of 75 m where a clay formation begins. No virus was found in observation wells at 6- and 9-m depth (Table 9.6). Low annual precipitation in this area may explain

TABLE 9.6. Monitoring of Wells for Viruses at Flushing Meadows, Arizona[a]

	Viruses per 100 Liters	
	Sewage Effluent	Well Water
A	786	0
B	2745	0
C	2378	0
D	158	0
E	7475	0
F	1142	0
Range	158–7475	0

[a] Adapted from R. G. Gilbert et al. (1976), *Science* **192**:1004–1005.

the failure to detect viruses (See Effect of Cations on Virus Adsorption below). In this project, viruses were concentrated by a relatively sophisticated concentration technique based on adsorption–elution concept (see Chapter 5) and capable of processing more than 100 gal groundwater.

In contrast, virus breakthrough was reported for two land application sites in Florida. In the St. Petersburg site, which received from 2 to 11 in./week of activated sludge effluent, viruses were isolated, after heavy rains, from 10- and 20-ft-deep wells.

A cypress dome, located in Gainesville, Florida, has been considered for the application of sewage effluents originating from a package treatment plant serving a mobile home community. Monitoring of viruses in observation wells has led to their demonstration in a 305-cm-deep well. The data suggested a vertical as well as a lateral movement of viruses. It was postulated, as in the St. Petersburg study, that heavy rains may have been responsible for the redistribution of viruses within the soil matrix and their subsequent transport to the groundwater. It was also suggested that the virus breakthrough may have occurred following the possible damage to the confining clay layer which resulted from the installation of an observation tower on the test site. Furthermore, laboratory experiments have shown that humic substances present in the standing dome water may also be responsible for virus movement through the dome soil. This will be discussed in more detail in the next section.

In Massachusetts, nonchlorinated sewage effluents from an Imhoff tank have been applied to a rapid infiltration site consisting of inconsolidated silty sands and gravels. The site was operated on the basis of 2–3-day wastewater application periods followed by 10-day resting periods. The presence of viruses was demonstrated in well water (4.2–10.6 PFU/l). A tracer study with bacterial phage f2 has been undertaken on the same test site. The phage was demonstrated in groundwater 48 hours after wastewater application (Figure 9.6).

This study has also shown the horizontal movement of viruses to distances greater than 180 m.

In Long Island wastewater chlorinated effluents were applied to recharge basins consisting of coarse sand. In the three locations under study the distance to the groundwater aquifer ranged from 30 to 80 feet. Enteroviruses, mainly echoviruses, were detected in observation wells at concentrations ranging from 1.3 to 10.6 PFU/gal. No virus was isolated however from a site where the water table was at a depth of 80 feet and where the soil consisted of coarse sand and fine silt.

From the above examples one can draw the following conclusions:

1. Viruses may contaminate groundwater following land disposal of sewage effluents.

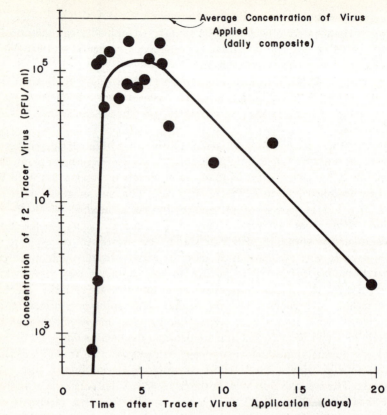

FIG. 9.6. Rapid infiltration site: Movement of bacterial phage f2 through soil treated with Imhoff tank effluents. From S. A. Schaub and C. A. Sorber (1977), *Appl. Environ. Microbiol.* **33:**609–619. Courtesy of American Society for Microbiology.

2. In irrigation sites the use of soils with adequate texture (e.g., fine loamy sand) should present no real hazard with respect to groundwater contamination.

3. In high-rate infiltration systems the use of very coarse soils (e.g., fine gravel and coarse sand) should be avoided. Furthermore the recharge basin should have an appreciable depth to the groundwater aquifer.

4. Rainwater plays a role in the redistribution of viruses within the profile of some soils.

5. In any virus monitoring program for land disposal sites, one should include the sampling of relatively large volumes of groundwater (e.g., 50 to 100 gal) and use adequate concentration techniques for virus detec-

tion. Soil from the disposal site should also be examined for the presence of viruses. Methodology for virus detection in soils is examined in Section 9.4.3.

Factors Controlling the Retention of Viruses by the Soil Matrix

The field studies have stimulated research on the mechanisms and the factors involved in virus removal by soils. Most of the early research efforts were focused on the removal of bacterial cells by soils and comparatively less is known about virus transport through soils. It is assumed that the transport of bacterial cells through soils results in their removal by

- *Sieving.* Bacterial removal increases when the particle size of the soil components decreases. Bacteria accumulating at the soil surface may in turn act as a filter for other bacteria and even smaller particles.
- *Sedimentation.* This occurs at the soil surface.
- *Adsorption.* This depends on many factors that will be discussed later for viruses.

The removal of bacteria in the top soil layer is illustrated in Table 9.7. Most of the fecal coliforms (90%) are concentrated in the top 13 cm of a fine sand.

TABLE 9.7. Vertical Distribution of Fecal Coliforms Through a Fine Sand Receiving Cow Manure Slurry[a,b]

Depth of Sampling (cm)	Number of cells/ml of Soil Solution	Number of cells/g Soil
3	110,000	350,000
13	24,000	28,000
28	4,300	5,200
38	1,500	2,500
48	<3	<3
58	<3	<4
93	<3	—

[a] Adapted from F. Dazzo et al. (1972), *Soil Crop Sci. Soc. Florida* **32**:99–102.

[b] The fresh cow slurry contained 2.4×10^5 fecal coliforms/ml.

We have so far seen that sieving can be an important removal mechanism for particles of the size of a bacteria. This mechanism becomes negligible when one considers the retention of colloidal size-particles such as viruses. Viruses adsorbed to large particles may however be retained by sieving and sedimentation processes. Therefore, studies on virus removal were mainly concentrated on the factors controlling the adsorption and transport of viruses to soils. Batch and column experiments have helped identify these factors. The major parameters are as follows:

- Type of soil.
- Flow rate of the percolating solution through the soil profile.
- Degree of saturation of pores in the soil.
- pH.
- Amount and type of cations in the percolating solution.
- Soluble organic materials.
- Type of virus.

We will now discuss the implication of each of these parameters in the removal of virus by soil systems.

Type of Soil. In our brief introduction to soils it was shown that among soil components clay minerals constitute an active fraction which is involved in the adsorption of biological and chemical pollutants by soils. This is mainly due to their great surface area and to their ion-exchange properties. Significant amounts of viruses are retained on clay surfaces and their adsorption depends on the type of clay under consideration. The virus–clay association is rapidly established and depends on the physicochemical characteristics of the suspending fluid (pH, cations, soluble organics). It is therefore assumed that the presence of clays in soils will result in an increased virus retention, and this is primarily due to an increase in surface area of the soil. Soils generally retain viruses more efficiently than pure sand. Figure 9.7 shows the adsorption of poliovirus to a sandy clay loam containing 28% of kaolinite. This soil displayed a good retention toward viruses and no breakthrough was observed even after application of eight pore volumes (43.7 cm) of groundwater.

Little is known about the virus sorptive ability of organic matter, the other important colloidal fraction of soils. Recent evidence suggests that organic soils are not as efficient as mineral soils for virus retention during application of wastewater effluents on land.

Iron oxides (magnetite, hematite, goetite) probably play a role in the adsorption of viruses to soils but their role remains to be determined.

From a recent study on the adsorption of several enteroviruses and bacteriophages to nine soils, it was concluded that no single soil can be used as a model for virus adsorptive capacity.

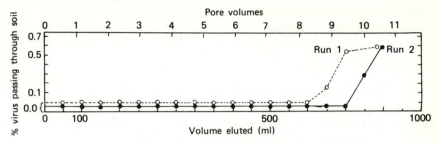

FIG. 9.7. Movement of poliovirus through a soil containing 28% kaolinite. From G. Bitton, P. R. Scheuerman, and A. R. Overman (unpublished).

Degree of Saturation of Pores in the Soil. The virus retention tends to decrease when the degree of saturation of the pores increases. Saturated pores may reduce chances for virus contact with soil particles. This has practical implications in the land disposal of sewage effluents. Soil drying between sewage applications may prevent the desorption of viruses from soil by rainwater. Table 9.8 illustrates the effect of the drying period on the ability of

TABLE 9.8. Effect of Drying Period on the Ability of Distilled Water to Desorb Soil-bound Viruses[a,b]

Soil Depth (cm)	PFU/ml			
	0.1[c]	1	3	5
2	50	5	0	0
5	8	25	0	0
10	650	35	0	0
20	128	0	20	0
40	328	0	20	0
80	145	0	0	0
160	0	0	0	0
240	0	0	0	0
250	0	0	0	0

[a] Adapted from J. C. Lance et al. (1976), *Appl. Environ. Microbiol.* **32:**520–526.

[b] The soil was flooded with deionized water for 1 day before sampling.

[c] Number of days of drying before the addition of deionized water.

distilled water (which simulates rainwater) to desorb soil-bound viruses. After 2 hours of drying, the distilled water was able to desorb some viruses. But when the drying period was increased from 2 hours to 5 days, no virus was desorbed from the soil.

Flow Rate. Most investigators have used sand to study the effect of flow rate on virus movement. Their results generally show that an increase in flow rate results in a decrease in viral adsorption. This information is particularly important in the performance of sand filters.

pH. Viruses may be viewed as electrically charged colloidal particles. The electrical charges are due to the presence of ionogenic groups (carboxyl and amino groups) on the protein coat. The pH of the suspending medium readily influence the ionization of the carboxyl and amino groups. Viruses behave as negatively charged particles at high pH values. Soil particles behave similarly and are also negatively charged on the basic side of the pH scale. Therefore, at high pH, viruses and soil particles repel each other and no adsorption occurs. As the pH is decreased below 7, virus adsorption to soils increases. Figure 9.8 illustrates the effect of pH on the adsorption of bacterial phage T1 to 5 different soils. The effect of pH appears to vary with the soil type. For some soils, the adsorption decreases to less than 40% when the pH = 8. A negative correlation is generally observed between soil pH and viral adsorption to soils.

Effect of Cations Virus Adsorption. The *concentration* and *species* of cation are important in the adsorption of viruses to soils. An increase in cation concentration brings about an increase in virus adsorption. Moreover, the extent of virus adsorption proceeds in the following way:

$$\text{Trivalent cation} > \text{Divalent cations} > \text{Monovalent cations}$$
$$(Al^{3+}, Fe^{3+}) \qquad (Mg^{2+}, Ca^{2+}) \qquad (Na^+, K^+)$$

Figure 9.9 shows the effect of cation concentration on virus adsorption to five soils. It appears that the cation concentration effect also depends on the soil under consideration. For example, soil 1 and 2 displayed a significant increase in virus (T2 phage) retention when cation concentration was increased.

Once viruses are adsorbed to soils, they may be released from the soil particle surface following a decrease in ionic strength of the percolating solution. This phenomenon probably operates under field conditions when rainwater percolates through the soil profile. Rainwater has a low ionic strength in the order of 20 μmhos/cm. This low value is close to the ionic strength of distilled water (around 2 μmhos/cm). Therefore, percolation of rainwater through soils will lead to a decrease in the ionic strength of the soil solution. Some investigators have observed a "burst" of viruses after a heavy rainfall on

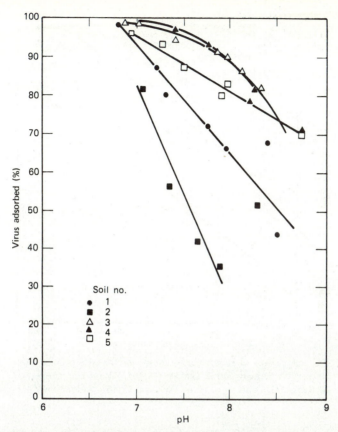

FIG. 9.8. Effect of pH on T1 adsorption to soils. From W. A. Drewry and R. Eliassen (1968), *J. Water Poll. Control Fed.* **40**:R257–271. Courtesy of Water Pollution Control Federation.

land disposal sites. This makes rainwater an important factor controlling the redistribution of viruses within the soil matrix. Laboratory experiments, using soil columns, have confirmed this important observation. After adsorbing viruses to soil, the columns are leached with distilled water (to simulate rainwater) and the viruses passing through the column are counted. It has been generally observed that viruses are easily desorbed from the soil. This phenomenon is well illustrated in Figure 9.10: Soil columns were alternatively treated with sewage effluents and distilled water. The breakthrough of poliovirus was monitored throughout three cycles. It is seen that maximum virus breakthrough corresponds to minimum *conductivity*. (Conductivity measures the ionic strength of a solution and is expressed in mmhos or μmhos/

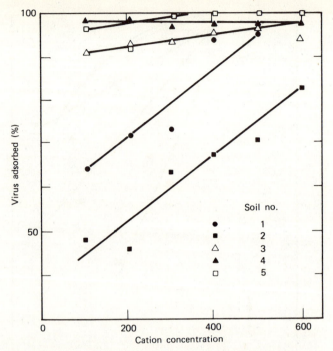

FIG. 9.9. Effect of cation concentration on T2 adsorption to soils. From W. A. Drewry and R. Eliassen (1968), *J. Water Poll. Contr. Fed.* **40:**R257–271. Courtesy of Water Pollution Control Federation.

cm at 25°C) Total organic carbon (TOC) breakthrough followed a pattern similar to viruses. However, virus desorption can be decreased or prevented by allowing the soil to dry for several days. It is important to point out that the adsorption–elution behavior of viruses is strain-dependent and that the transport pattern of enteric viruses cannot be predicted accurately.

Effect of Soluble Organic Substances on the Adsorption Process. It is known that soluble organic materials can lower the adsorption of viruses to solid surfaces. This phenomenon may also be valid for virus–soil interaction. Proteins present in sewage effluents may theoretically compete with viruses for the adsorption sites on soils. However, sewage effluents do not hinder in a significant way the attachment of viruses to soils. An activated sludge effluent may have a conductivity ranging from 500 to 700 μmhos/cm. This ionic strength might be a factor leading to enhanced virus adsorption. Poliovirus, suspended in activated sludge effluent, was found to be efficiently retained by a sandy soil. Only 0.008% of applied virus passed through the soil columns after percolation of 20 pore volumes of rainwater. Thus, organics present in wastewater effluents do not seem to have a significant effect on virus transport through soils.

Humic substances are another category of organic materials which may have an influence on the transport pattern of viruses through soils. These substances give a brown-yellow color to natural waters. They are the product of microbial decomposition of plant materials. They have been classified into four main categories: humin, and humic, hymatomelanic, and fulvic acids. Fulvic acid, having the least complex structure, is regarded as the dominant form in freshwater habitats. There is a particular situation where these substances may influence the retention of viruses by soils. This situation is encountered when wetlands are used as a disposal site for wastewater effluents. This disposal practice was studied at the University of Florida: Secondary effluents entering a cypress dome become heavily colored with "humic substances" resulting from

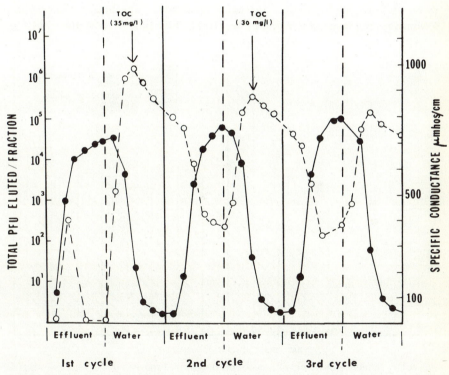

FIG. 9.10.   Effect of ionic strength of percolating water on adsorption of poliovirus 1 to soil. From S. M. Duboise et al. (1976), *Appl. Environ. Microbiol.* **31:**536–543. Courtesy of American Society for Microbiology. O ---O PFU, ●——● Specific conductance.

the decomposition of cypress needles. Laboratory experiments have shown that these materials readily interfere with the adsorption of viruses to the soil matrix. Figure 9.11 shows the movement of poliovirus through a sandy soil. The virus was readily adsorbed in the presence of tap water but not in the presence of colored water (420–750 color units). Sediment leachates with a high color content (2000 color units) allow the passage of approximately 80% of the virus applied to sandy soil columns. The treatment of these leachates with activated carbon to remove color restores the adsorption capacity of the soil (Figure 9.12). It was recently demonstrated that fulvic acids may actually complex viruses and prevent their adsorption to soils.

Type of Virus. Once we have a knowledge of the various factors which govern virus retention by the soil matrix, one may be tempted to draw some general conclusions concerning virus transport through soils. However, one must proceed with caution and remember that there may be some subtle differences in the surface characteristics of viruses. T phages (e.g., T2, T4, T7) were used as model viruses in many laboratory experiments. These bacteriophages are large in size and possess a tail. It is difficult to compare them to

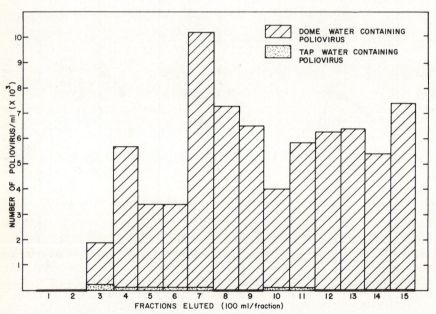

FIG. 9.11. Influence of humic substances on poliovirus 1 movement through a sandy soil. From G. Bitton et al. (1976), *J. Environ. Qual.* **5**:370–375. Courtesy of Amer. Soc. Agron., Crop Science Society of America and Soil Science Society of America.

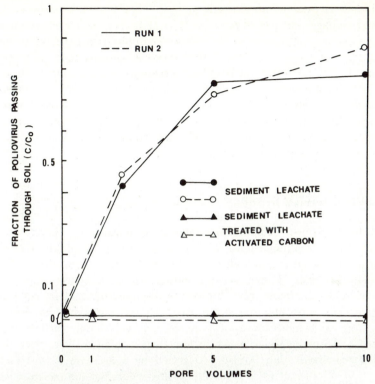

FIG. 9.12.   Effect of sediment leachates on the movement of poliovirus 1 through a brown red sandy soil. From P. R. Scheuerman et al. (1979), *J. Env. Eng. Div.* **105**:629–640. Courtesy of American Society of Civil Engineers.

small tailless picornaviruses. Therefore, the virus used must be as close as possible to the actual enteric viruses. The use of RNA tailless phages such as MS2 or f2 probably should be a better alternative. However, recent studies have revealed that f2 adsorbs poorly to soils and thus does not appear to be a suitable model for the study of the sorptive pattern of enteroviruses. Enteroviruses also differ in their adsorption to soils. Recent laboratory experiments have shown that among the many enteroviruses examined echovirus 1 was the least adsorbed to a sandy soil. It also appears that adsorption to soil is strain-dependent. Interestingly, echoviruses were the predominant viruses found in wells from recharge basins treated with sewage effluents.

Laboratory studies (batch and column experiments) have significantly increased our understanding of virus transport through soils under field conditions. They have revealed that rainwater is an important factor in the redistribution of viruses within the soil matrix. They have also suggested some potential management practices to avoid viral contamination of groundwater. For example, soil drying between wastewater application periods can help in the retention of viruses by soils. Laboratory experiments have confirmed the findings of field studies on, at least, two occasions. Unfortunately, there is no standard soil column test to assess the potential of soils to retain viruses. This test should be involved in the consideration of a site for the disposal of sewage effluents. This test, in combination with other parameters (e.g., depth to the water table, regime of wastewater application, climatological data) should be of great use in the selection of land disposal sites.

9.4.3   Virus Survival in Soils

A realistic assessment of potential public health hazards associated with land disposal of sewage effluents must take into consideration the ability of viruses to survive in the soil. It is now well established that viruses retain their infectivity in the adsorbed state. They survive in soils for up to six months, depending on environmental conditions prevailing at the disposal site. Soil microflora does not appear to exert any significant effect on virus survival. However, soil temperature and moisture appear to be the controlling factors in viral persistence. Viruses survive better at low than at high temperatures (Figure 9.13). Virus inactivation is also accelerated by a decrease in soil moisture (Table 9.9). Virus survival is generally greatly reduced at soil moisture content below 10%. Evaporation of soil water may be the main factor responsible for virus inactivation in drying soils. Through the use of radioactively-labeled viruses, it has been demonstrated that viruses are truly inactivated rather than irreversibly attached to dry soils. Inactivation is the result of dissociation of virus components as well as the degradation of the nucleic acid core.

The assessment of virus survival in soils treated with sewage effluents or sludge is possible only through the development of sensitive and rapid techniques for the recovery of viruses from soils. An ideal method for detecting low numbers of viruses in soil systems should allow an efficient elution of viruses from the soil matrix and provide a means for concentration of soil eluates in order to achieve an efficient recovery of low numbers of viruses. Three eluents have been evaluated for their ability to recover viruses from soils. These eluents are 0.25M glycine in 0.05M EDTA (pH = 11.0), 3% beef extract (pH = 9.0) and 0.5% isoelectric casein (pH = 9.0). The overall virus recovery achieved ranged from 37% with beef extract to 60% with isoelectric casein (Table 9.10). A general scheme for virus recovery from soil by the isoelectric casein method is outlined in Figure 9.14.

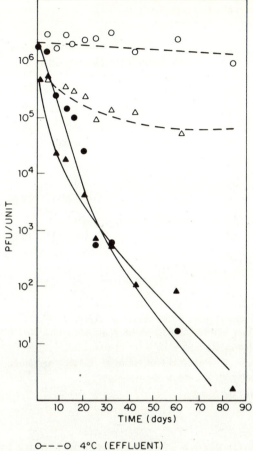

FIG. 9.13. Effect of temperature on survival of poliovirus in soil cores and sewage effluents. From S. M. Duboise, et al. (1976), *Appl. Environ. Microbiol.* **31**:536–543. Courtesy of American Society for Microbiology.

227

TABLE 9.9. Effect of Soil Moisture on the Survival of Poliovirus 1 at 20°C[a]

	% Virus Recovered		
Days	25% Moisture	Drying Conditions[b]	
1	69	74	(13.1)
3	41	35	(10.9)
8	22	0.3	(6.2)
10	17	0.08	(5.0)
14	13	0.02	(4.6)
21	10	ND[c]	
28	5	0.003	(4.6)
42	2	ND	
49	1	0.002	(4.6)
80	0.2	ND	
100	0.07	ND	
134	0.004	ND	

[a] Adapted from B. Moore et al. (1976), *Proc. 3rd Natl. Conf. Sludge Management, Disposal and Utilization,* Miami, Fla.

[b] Soil allowed to dry at room temperature. Values in parentheses are % soil moisture.

[c] Not detected.

TABLE 9.10. Overall Recovery of Poliovirus Type 1 from a Sandy Soil Using Three Different Methods[a]

Recovery Method	Total Virus Input (PFU)	% Elution	Concentration Step % Recovery	Overall Recovery %
3% beef extract (pH = 9.0)	5.0×10^5	48	78	37
0.25 M Glycine + 0.05 M EDTA (pH = 11.0)	3.8×10^5	61	67	41
0.5% Isoelectric casein (pH = 9.0)	4.7×10^5	126	48	60

[a] Adapted from G. Bitton, J. Charles, and S. R. Farrah (1979), *Can. J. Microbiol.* **25**:874–880.

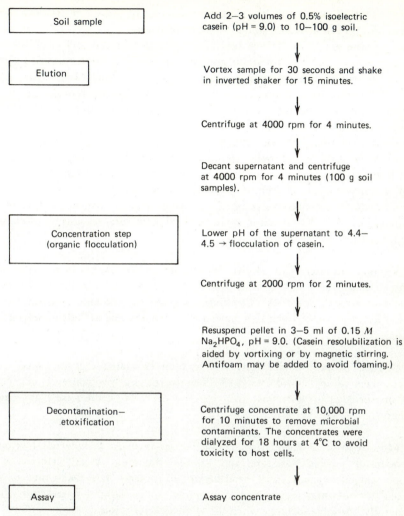

Soil sample	Add 2–3 volumes of 0.5% isoelectric casein (pH = 9.0) to 10–100 g soil.
Elution	Vortex sample for 30 seconds and shake in inverted shaker for 15 minutes.
	Centrifuge at 4000 rpm for 4 minutes.
	Decant supernatant and centrifuge at 4000 rpm for 4 minutes (100 g soil samples).
Concentration step (organic flocculation)	Lower pH of the supernatant to 4.4–4.5 → flocculation of casein.
	Centrifuge at 2000 rpm for 2 minutes.
	Resuspend pellet in 3–5 ml of 0.15 M Na_2HPO_4, pH = 9.0. (Casein resolubilization is aided by vortixing or by magnetic stirring. Antifoam may be added to avoid foaming.)
Decontamination— etoxification	Centrifuge concentrate at 10,000 rpm for 10 minutes to remove microbial contaminants. The concentrates were dialyzed for 18 hours at 4°C to avoid toxicity to host cells.
Assay	Assay concentrate

FIG. 9.14.   General scheme for virus recovery from soil by the isoelectric casein method. From G. Bitton, M. J. Charles, and S. R. Farrah, (1979), *Can. J. Microb.* **25:**874–880. Courtesy of National Research Council of Canada.

9.5   PRODUCTION OF MICROBIAL AEROSOLS FOLLOWING SPRAY IRRIGATION WITH WASTEWATER EFFLUENTS

We have seen that raw municipal wastewater contains viruses ranging in number from 7×10^3 to more than 10^5 PFU/l. We know that modern wastewater treatment processes do not completely remove the virus threat. Any mechanical device which assists in the aerosolization of raw wastewater or wastewater effluents may lead to the production of biological aerosols. Thus, aeration tanks in activated sludge systems, trickling filters, and spray irrigation with wastewater effluents are typical sources of bacterial and viral aerosols. These aerosols may remain viable and be carried over great distances. Bacterial aerosols produced by wastewater treatment plants have been detected at distances greater than 1 km. The survival of aerosolized microorganisms is primarily controlled by meteorological factors, namely relative humidity, solar radiation, temperature, and wind velocity.

Spray irrigation with wastewater effluents has been in use for some time in many parts of the world. Some countries have issued guidelines which regulate the practice of wastewater treatment by spray irrigation and sometimes require buffer zones to protect human populations. Within the United States some states require chlorination prior to spraying of crops. An evaluation of the potential hazard due to viral aerosols must take into account the following important factors.

Virus Concentration of the Wastewater. We have seen that the virus load of wastewater was variable and was not significantly reduced by conventional treatment processes. Therefore, wastewater effluents are suitable for spray irrigation only when they have a low turbidity and are adequately chlorinated.

Degree of Aerosolization. It has been estimated that the wastewater aerosolized varies from 0.2 to 0.4%. The degree of aerosolization is dependent on meteorological conditions, and type and operating conditions of the equipment.

Meteorological Conditions. The factors involved include wind velocity, temperature, solar radiation and relative humidity. Wind velocity is an important factor controlling the transport of aerosols. It is taken into account when formulating regulations concerning spray irrigation practice. It is generally advisable to spray when the wind velocity is low.

9.5.1 Survival of Viral Aerosols

The persistence of viruses and bacteria in the airborne state depends upon many environmental factors, the most important of which being relative humidity (RH), desiccation, solar radiation and temperature. The decline of microbes in the airborne state proceeds in two stages.

- *First stage.* There is a rapid die-off of the microbe following the initial shock due to desiccation. This stage lasts some seconds and it has been estimated that $0.5 \log_{10}$ of microbes undergo inactivation.

- *Second stage.* This stage is slower and is controlled by the factors mentioned above, namely relative humidity, temperature, and solar radiation.

Since the effect of temperature and solar radiation was discussed in Chapters 3 and 4, one need only examine here the impact of RH on airborne viruses. Nonenveloped viruses (picornaviruses and adenoviruses) survive best at high relative humidity, whereas enveloped viruses (e.g., myxoviruses, paramyxoviruses, Semliki forest virus) survive best at low relative humidity. Figure 9.15 illustrates this phemomenon with regard to poliovirus and influenza virus. The relatively increased survival of enteric viruses at high RH has also been confirmed for some coxsackie- and echoviruses. Once in the airborne state, viruses may be inactivated by exposure to the air–water interface or by desiccation (see Chapter 3). The protein coat appears to be the target for inactivation of some aerosolized viruses, namely poliovirus. The RNA core retains its infectivity following the aerosolization process (Figure 9.16). The epidemiological significance of this phenomenon is questionable since the infectivity of RNA is approximately 0.1% of that of intact viruses.

9.5.2 Health Aspects of Spray Irrigation with Wastewater Effluents

Many studies have demonstrated the production of viable aerosols by various sources including spray irrigation with wastewater effluents. Unfortunately, these field studies have dealt mainly with the sampling of bacterial or bacteriophage aerosols. An airborne enterovirus, echovirus 7, was detected at 40 m downwind from a sprinkler in a spray irrigation site in Israel. At Pleasanton, California, enteroviruses were also detected at 50 m downwind from the

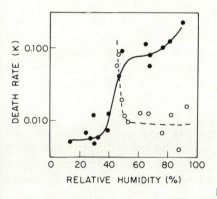

INFLUENZA VIRUS
POLIOVIRUS

FIG. 9.15. Influence of relative humidity on the survival of influenza virus and poliovirus. From J. H. Hemmes, et al. (1960), *Nature* **188**:430. From Courtesy of *Nature*.

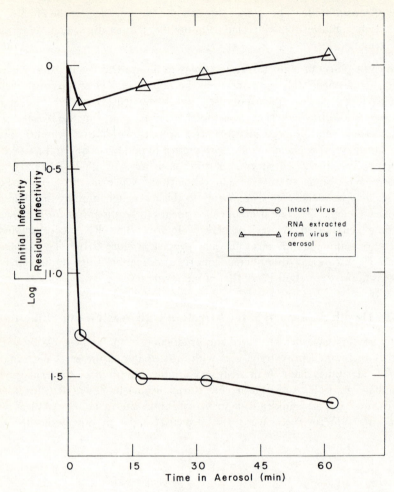

FIG. 9.16.   Inactivation of the infectivity of poliovirus and its RNA in aerosols at 35% relative humidity. Adapted from J. C. DeJong et al. (1973), *J. Gen. Virol.* **18**:83–86.

sprinklers. In this particular site the sewage effluent was not chlorinated and was sprayed at a rate of 1.5 mgd on grassland. The calculation of aerosol:wastewater ratio has led to the conclusion that airborne enteroviruses are hardier than bacteria or coliphages. Unfortunately, monitoring of airborne enteroviruses appears complicated and necessitates sampling of large volumes of air with the aid of costly high-volume samplers (1000 liter air/minute). In view of these difficulties, prediction models appear therefore necessary to estimate airborne virus concentrations. These models are based on prior knowledge of virus concentration in the wastewater effluent being sprayed.

Apart from influenza virus and other respiratory pathogens, some enteric viruses are probably transmitted via air to man and animals. The possibility of airborne transmission of infectious hepatitis virus and even oncogenic viruses has been considered. As it is the case for waterborne transmission one must take into account the *infective dose* of the pathogen. We have seen that 1 PFU may represent an infectious dose for viruses. For bacteria, the infectious dose is much higher. However, one is compelled to report a disturbing finding in experiments using animals (chimpanzees); the dose of airborne *Salmonella typhi* that caused disease in the animals was 1000 times lower that the one necessary to cause the disease via the oral route. The size of the airborne particles is another factor that one must consider when attempting to assess the public hazard due to airborne microbes. Due to their small size (0.02 to 0.2 μm) viruses fall in the category of airborne particles that are effectively retained at the level of the alveoli in the lungs.

Particle size determines the degree of retention in the respiratory system and the depth of penetration before deposition. Particles with a size range of 1–2 μm escape trapping by the upper respiratory tract and display the greatest deposition at the level of the alveoli. The alveolar deposition decreases when aerosols have a size between 1 and 0.25 μm and then increases again when the particle size is below 0.25 μm. This increase is due to Brownian motion.

The health effects of microbial aerosols concern primarily the sewage treatment plant workers and the nearby population which are chronically exposed to this hazard. Some preliminary studies did not show any health effect. It has been suggested that this segment of the population builds up immunity, since they are chronically exposed to low concentrations of airborne microbes.

With regard to viruses, the possibility of health hazard has been derived from dispersion models. In any study on health aspects of spray irrigation it is desirable to proceed according to the following:

1. *Sampling of Viruses in spray irrigation site.* Researchers at the University of Texas in San Antonio have recently developed a method for the recovery of airborne viruses at the spray irrigation site: 30 m³ air (1000 liters/min for 30 min) are collected in 100 ml of brain heart infusion broth containing 0.1% Tween 80. The collection fluid is further concentrated by the two-phase separation technique (see Chapter 5).

2. *Epidemiological studies.* These are tedious and expensive. They must correlate disease with the occurence of viruses in the field. One retrospective study dealt with the risk of disease infection associated with spray irrigation in agricultural settlements in Israel. The study compared 77 kibbutzim (a *kibbutz* is a cooperative agricultural settlement)

that use spray irrigation with nonchlorinated oxidation pond effluents to 130 others that do not use it. The diseases investigated were shigellosis, salmonellosis, infectious hepatitis (all three are waterborne), influenza (airborne), streptococcal infection and tuberculosis. Figure 9.17 shows that, during the irrigation season (April–November), there was a higher incidence of diseases in kibbutzim using spray irrigation, whereas during the winter, no difference was observed in the incidence of infectious hepatitis and shigellosis.

In conclusion, it appears that the use of nonchlorinated sewage effluents may result in an increase of the incidence of enteric diseases among populations in close proximity to spray irrigation sites. Prior to spray irrigation with sewage effluents one must thus establish an adequate buffer zone to protect neighboring populations and/or efficiently chlorinate sewage effluents.

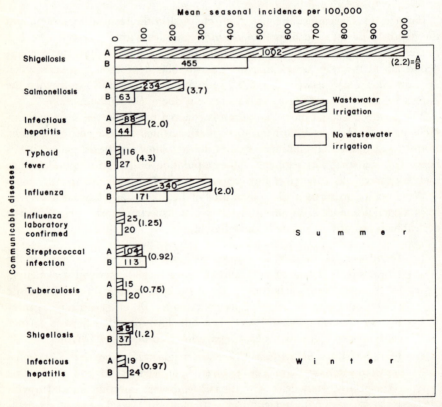

FIG. 9.17. Risk of communicable disease infection associated with wastewater irrigation in agricultural settlements in Israel. From E. Katzenelson et al. (1976), *Science*, **194**:944–946. Copyright 1976 by the American Association for the Advancement of Science.

9.5.3   Prediction of Virus Concentration in Wastewater Aerosols

Dispersion models, borrowed from the field of aerosol mechanics, have been proposed to predict the downwind levels of aerosolized microbes. Unfortunately, these models make too many assumptions (e.g., no microbial die-off) and give higher numbers of microbes than those found in the field. Recently, a more sophisticated model has been proposed to predict the microbiological aerosol concentrations in a spray irrigation site, in Pleasanton, California. This model incorporates essential microbiological factors such as decay rates.

Given a spray irrigation site, the predicted concentration P of microbial aerosols downward from a sprinkler is given by the equation:

$$P = D \times E \times I \times e^{\lambda(a - a_{50})} \tag{10.1}$$

where $D = D_g(r,d)$ = physical diffusion model aerosol concentration of microorganism g emanating from the spray line to a distance d during run r, assuming all the sprayed wastewater (including its measured wastewater concentration of microorganism g) becomes aerosol, and assuming no microbiological die-off (CFU/m^3).

D depends on many factors including microbial concentration in the wastewater effluent being sprayed, meteorological conditions, topography, and sprinkler configuration.

E = *aerosolization efficiency factor,* that is the fraction of wastewater effluent which is aerosolized ($O < E < 1$). This factor depends on temperature, wind velocity and solar radiation. This factor is therefore site-specific.

I = *microbiological impact factor*: It is the proportion of aerosolized microorganisms that remain viable at a distance, $d = 50$ m ($I > O$)

λ = microbiological age decay rate

a = aerosol age

This model has been evaluated for the Pleasanton site and the results are illustrated in Table 9.11.

It has been suggested that this prediction model can help avoid the expenses and the problems encountered during sampling of viral aerosols. If one opts for air sampling, the model can help predict the volume of air to be sampled.

9.6   SEPTIC TANKS AND SANITARY LANDFILLS

These systems constitute a way to dispose of household wastes and solid wastes, respectively. They are mentioned in this chapter because they involve soil as an adsorption matrix.

TABLE 9.11. Prediction of Nighttime Microorganism Aerosol Levels Entering Pleasanton Residential Area[a]

	Summer Nighttime Case
Model Input Conditions	
Season and time	Summer night
Air temperature	20° C
Relative humidity	70 %
Solar radiation	0 W/m²
Wind velocity	2 m/s
Stability class	E
Mixing height	30 m
Residential distance	650 m
Residential direction	E to SE
Aerosol Age, a	325 s
Wastewater spray rate	70 l/s

	Microorganism Group		
	Total Coliform	Mycobacteria	Enteroviruses
Wastewater concentration	1×10^7 MFC/l	80,000 CFU/l	50 PFU/l

Model Parameter Values

D (centerline)	12,000 MFC/m³	100 CFU/m³	0.06 PFU/m³
E	0.0033	0.0033	0.0033
I	0.17	4.3	100
λ	-0.011 s^{-1}	-0.006^{b} s^{-1}	-0.002^{b} s^{-1}
$e^{\lambda(a - a_{50})}$	0.037	0.17	0.55

Predicted Aerosol Concentration

P (centerline)	0.2 MFC/m³	0.2 CFU/m³	0.01 PFU/m³

[a] D. E. Camann et al. (1978), *in: Risk Assessment and Health Effects of Land Application of Municipal Wastewater and Sludges*, B. P. Sagik and C. A. Sorber, Eds. Center for Applied Research and Technology, University of Texas, San Antonio.

[b] Interpolated value beyond the quantified range of the model parameter.

236

9.6.1   Septic Tanks

In the United States approximately 25% of the population is still served by septic tanks. The house sewage enters a *septic tank* and the effluent produced is finally discharged into an *absorption field* (Figure 9.18):

Septic tank. The tank is made of concrete, metal, or fiberglass and is designed to remove the sewage solids to avoid clogging of the absorption field. The sewage undergoes anaerobic digestion, which results in the production of *sludge* and of a floating layer of light solids (fats) called *scum*. The detention time of wastewater within a septic tank varies from 24 to 72 hours. These tanks should be inspected regularly and are cleaned every 3 to 5 years.

Absorption field. The effluents from septic tanks reach an absorption field through a system of perforated pipes which are surrounded by gravel or crushed stones. The soil used for the disposal of septic tank effluents is

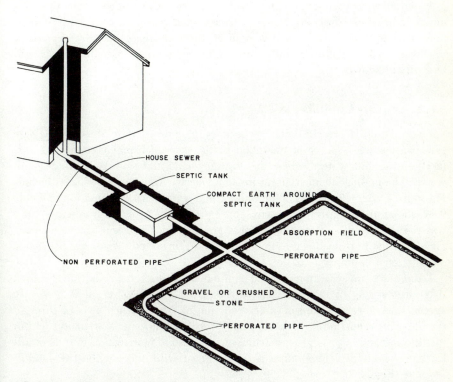

FIG. 9.18.   Septic tank sewage disposal system. Adapted from U.S. Dept. H.E.W. 1969. *Manual of Septic Tank Practices*, Public Health Serv. Publ. No. 528.

generally selected according to certain characteristics, namely, soil texture, structure, depth of permeable strata, types of clay minerals present.

It is well known that septic tanks effluents contain chemical and biological pollutants. Contamination of wells, soil, and surface water by septic tanks effluents may result in serious health hazards and in the degradation of our environment. Chemical pollutants found in these effluents may enter bodies of water and cause excessive growth of algae and aquatic weeds.

Since the sewage digestion taking place in a septic tank does not result in the complete inactivation of bacteria and viruses, one must be concerned about the fate of these pathogens in the absorption field. The transport of viruses through the soil is governed by the same factors we discussed earlier. The U.S. Public Health Service has recommended that a septic tank should be at least 100 feet from any water supply well and 50 feet from any surface water. Some states consider a soil marginally suitable for the disposal of septic tank effluents when it has a high fluctuating water table and impermeable layers which may lead to the surfacing of untreated or partially treated sewage effluents.

Laboratory experiments using soil columns have shown that properly designed systems effectively remove viruses from septic tank effluents. Under normal conditions, viruses will remain in the soil where they are eventually inactivated. The pollution of groundwater by septic tank effluents is still basically unknown.

9.6.2 Sanitary Landfills

Sanitary landfills are the major repository of municipal solid wastes. These solid wastes are a source of microbial pathogens which may contaminate sur- face- and groundwaters when the sanitary landfill becomes saturated with water. The pathogenic microorganisms may derive from various sources includ- ing soiled hygienic products, food, and animal wastes. An examination of Table 9.12 shows that diapers represent approximately 3% of the municipal solid waste. It has been reported that diapers are the major source of viruses in sani- tary landfills. For obvious reasons, the three strains of poliovirus are the major ones detected in diapers.

Laboratory investigations, using simulated sanitary landfills, have shown that viruses may or may not pass through the fill. However, it is generally agreed that there is a considerable reduction of viruses when passing through the fill. This reduction is probably due to restricted movement of viruses within the fill (adsorption to solid waste components) and to inactivation in the landfill leachate. Antiviral effect appears to increase with the age of the landfill leachate (Figure 9.19).

The surviving viruses passing through the fill may bring about a contamina- tion of surface- and groundwater. Their transport to the groundwater supply is

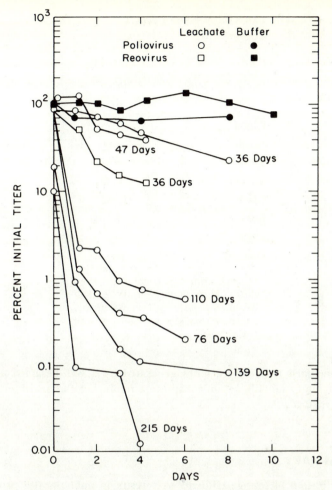

FIG. 9.19. Survival of enteric viruses in sanitary landfill leachates collected after various days of lysimeter operation. From R. S. Engelbrecht et al. (1974), in: *Virus Survival in Water and Wastewater Systems,* J. F. Malina Jr. and B. P. Sagik, Eds. The University of Texas, Austin. Courtesy of Center for Research in Water Resources, University of Texas, Austin.

controlled by factors which we discussed previously. A survey was recently undertaken by the University of North Carolina on the occurrence of enteric viruses in sanitary landfill leachates under field conditions. Only one out of 22 leachates examined contained viruses (poliovirus type 1). This sample originated from a site with no adequate sanitary landfill practices.

TABLE 9.12.   Solid Waste Components[a]

Item	Percent
Food waste	11.63
Garden waste	8.34
Paper	43.75
Plastics	4.66
Wood	0.61
Metals	10.85
Glass	15.82
Rock	1.20
Rags	—
Diapers	3.13

[a] Adapted from R. S. Engelbrecht et al. (1974), *Virus Survival in Water and Wastewater Systems*. J. F. Malina, Jr., and B. P. Sagik, Eds., University of Texas, Austin.

It is concluded that the leachates generated by a well-designed and well-operated sanitary landfill do not constitute a public health hazard with regard to enteric viruses.

9.7 SUMMARY

Land application of sewage effluents may result in public health problems of microbial (viruses, bacteria) and chemical (nitrates, heavy metals) origin.

Field studies have shown that viruses may contaminate groundwater following land application of sewage effluents. Virus transport is controlled by the soil type, depth of the groundwater aquifer, and method of land application.

Laboratory experiments have shown that virus transport through soils is controlled by soil type, flow rate, degree of saturation of pores in the soil, pH, cations, soluble organic materials, and virus type. Rainwater is involved in the redistribution of viruses within the soil matrix.

Viruses may survive for long periods in soils. The survival pattern is controlled primarily by temperature and soil moisture.

Spray irrigation with wastewater effluents may result in the production of biological aerosols. Virus survival in the airborne state is influenced by temperature, solar radiation, and relative humidity. Wind velocity controls significantly the transport of biological aerosols, including viruses.

Little is known about the health aspects of viral aerosols originating from spray irrigation sites.

Properly designed septic tank systems effectively remove viruses from tank effluents.

Leachates generated by a well-designed and well-operated sanitary landfill do not constitute a public health hazard with regard to viruses.

9.8   FURTHER READING

Baldwin, L. B., J. M. Davidson, and J. Gerber, Eds. 1976. *Virus Aspects of Applying Municipal Wastes to Land*. University of Florida, Gainesville, Fla.

Benarde, M. A. 1973. Land disposal of sewage effluents: appraisal of health effects of pathogenic organisms. *J. Am. Water Works Assoc.* **65**:432–440.

Bitton, G. 1975. Adsorption of viruses to surfaces in soil and water. *Water Res.* **9**:473–484.

Bitton, G. 1980. Adsorption of viruses to surfaces: Technological and ecological implications. In: *Adsorption of Microorganisms to Surfaces*, G. Bitton and K. C. Marshall, Eds., Wiley New York.

Bouwer, H. 1976. Use of the earth's crust for treatment or storage of sewage effluent and other waste fluids. *Crit. Rev. Environ. Control* **6**:111.

Burge, W. D., and P. B. March. 1978. Infectious disease hazards of landspreading sewage wastes. *J. Environ. Quality* **7**:1–9.

Donahue, R. L., J. C. Shicklune, and L. S. Robertson. 1971. *Soils: An Introduction to Soils and Plant Growth*, Prentice-Hall, Englewood Cliffs, N.J.

Engelbrecht, R. S., and P. Amirhor. 1976. Disposal of municipal solid waste by sanitary landfill. In: *Virus Aspects of Applying Municipal Waste to Land*. L. B. Baldwin et al., Eds. University of Florida, Gainesville, Fla.

Gerba, C. P., C. Wallis, and J. L. Melnick. 1975. Fate of wastewater bacteria and viruses in soil. *J. Irrig. Drainage Div.* **101**:157–174.

Green, K. M. 1976. Sand filtration for virus purification of septic tank effluents. Ph.D. thesis. University of Wisconsin, Madison.

Sagik, B. P., and C. A. Sorber, Eds. 1978. *Risk Assessment and Health Effects of Land Application of Municipal Wastewater and Sludge*. Center for Applied Research and Technology, University of Texas at San Antonio, Texas.

Sorber, C. A. 1976. Virus in aerosolized wastewater. In: *Virus Aspects of Applying Municipal Wastes to Land.* L. B. Baldwin, et al., Eds. University of Florida, Gainesville, Fla.

Wellings, F. M., A. L. Lewis, and C. W. Mountain, 1977. Survival of viruses in soil under natural conditions. In: *Wastewater Renovation and Re-use.* F. M. D'Itri, Ed. Marcel Dekker, New York.

ten

Viruses in Sludges

10.1 INTRODUCTION

Sludge is a complex mixture of solids of biological and mineral origin that are removed from wastewater in sewage treatment plants. Sludge is a by-product of physical (primary treatment), biological (activated sludge, trickling filters) and physicochemical (chemical precipitation with lime, ferric chloride, or alum) treatment of wastewater. Approximately 5 millions dry tons of sludge were generated in 1976 by more than 20,000 municipal sewage treatment plants across the United States. This amount is expected to double by the time most of the sewage treatment plants pass to secondary treatment, thus magnifying the complex sludge management problem. Current disposal practices are further challenged by federal and state regulations (e.g., phasing out of ocean disposal) or by higher energy costs (e.g., incineration). A safe and cost effective solution must be found to the sludge problem.

Viruses associated with human and animal feces enter municipal wastewater, and some are merely transferred to sludge following their passage through sewage treatment plants.

In this chapter our purpose is to examine the fate of viruses in sludge treatment operations and to discuss the problems encountered in sludge disposal practices.

10.2   SLUDGE PROCESSING IN SEWAGE TREATMENT PLANTS

Sludge resulting from various stages of a sewage treatment plant (primary treatment, activated sludge, trickling filters) must be processed in order to accomplish the following goals:

- *Reduction of organic matter and water content.* This results in a weight reduction, which makes the transportation of sludge more economical.
- *Hygienic aspects.* The sludge must be rendered free of microbial pathogens and parasites.
- *Improvement of the aesthetic aspect of sludge by elimination of odors.*

Sludge processing involves 6 typical steps which we will briefly describe (Figure 10.1).

10.2.1   Thickening

The purpose of this step is to reduce the volume of the sludge. The solid concentration in a typical sludge is approximately 2% (w/v) and may be as high as 10% after thickening. Table 10.1 shows the solid concentration of three different sludges before and after thickening.

10.2.2   Digestion

This step involves microbiological processes that result in the stabilization of organic matter, gas production, and some destruction of microbial pathogens.

Anaerobic sludge digestion is carried out as a single stage or as a two-stage process and is described in Figure 10.2.

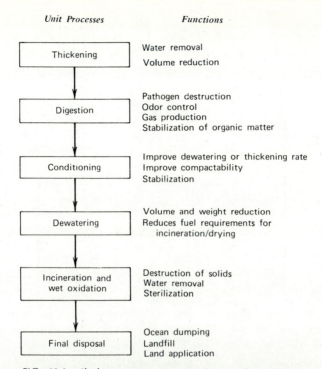

Unit Processes   *Functions*

| Thickening | Water removal
Volume reduction |

| Digestion | Pathogen destruction
Odor control
Gas production
Stabilization of organic matter |

| Conditioning | Improve dewatering or thickening rate
Improve compactability
Stabilization |

| Dewatering | Volume and weight reduction
Reduces fuel requirements for
    incineration/drying |

| Incineration and
wet oxidation | Destruction of solids
Water removal
Sterilization |

| Final disposal | Ocean dumping
Landfill
Land application |

FIG. 10.1. Sludge treatment processes and their function. Adapted from *Process Design Manual for Sludge Treatment and Disposal,* U.S. Environ. Prot. Agency Technology Transfer, EPA 625/1-74-006 (1974).

TABLE 10.1. Sludge Thickening[a]

	% Solids	
Sludge Type	*Before Thickening*	*After Thickening*
Primary	2.5–5.5	8–10
Trickling filter	4–7	7–9
Activated	0.5–1.2	2.5–3.3

[a] Adapted from Metcalf and Eddy Inc. (1972), *Wastewater Engineering,* McGraw-Hill Book Co., New York.

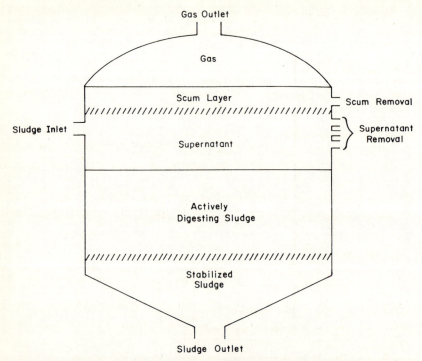

FIG. 10.2: Anaerobic digestion of sludge (Standard rate digestion). From U.S. Env. Prot. Agency, 1974. *Process Design Manual for Sludge Treatment and Disposal.* EPA 625/1-74-006. Courtesy of Environmental Protection Agency.

Two groups of bacteria participate in the anaerobic digestion of sludge: complex organics in the sludge are converted to volatile acids by bacteria called *acid formers*. The organic acids are then transformed to methane, carbon dioxide and other gases by *methanogenic bacteria*. The methane produced has a heating value of approximately 500–900 Btu/ft³ and may be used as an energy source in the treatment plant.

The optimal conditions for the anaerobic digestion are the following:

- pH ~ 7.
- Absence of oxygen.
- Temperature around 90–95°F (32–35°C).
- Sufficient alkalinity to neutralize the acidity produced by acid formers.
- Absence of toxic chemicals such as heavy metals, excess ammonia, or sulfides.

In some relatively small sewage treatment plants, sludge is digested aerobically. This process has the advantage of producing a sludge which can be dewatered more easily. However, it has two main drawbacks: additional cost associated with oxygen supply and absence of methane production.

10.2.3 Conditioning

The purpose of this operation is to improve the dewatering characteristics of the sludge. This goal can be achieved by using one of two methods:

- *Addition of Chemicals.* (Such as alum, lime, ferric chloride, or synthetic polyelectrolytes.) The latter aggregate sludge particles via interparticle bridging and charge neutralization. Cationic polymers work most efficiently at pH <7.
- *Heat Treatment.* (Under pressure.) This may also reduce the affinity of sludge for water.

10.2.4 Dewatering

Dewatering is accomplished by employing one of the following procedures:

Air-Drying. The sludge is spread on porous *drying beds* which are made of approximately 6 in. of sand supported by a gravel layer. The loss of water is accomplished through evaporation (which depends on atmospheric conditions) and percolation to an underdrain system. Air-drying can raise the solid content of the sludge to approximately 40% within 10 to 15 days.

Centrifugation. Large centrifuges are used to bring sludge solid content to approximately 60%. This operation does not generally require any conditioning.

Pressure Filtration. Chemically conditioned sludge is treated under pressure for 1 to 3 hours. The resulting sludge has a solid content between 30 and 45%.

Vacuum Filtration. Vacuum filtration is a widely used method for sludge dewatering. The solid content of the finished product ranges between 20 and 30%.

10.2.5 Incineration and Wet Combustion

Incineration is a dry combustion process which results in mineral ashes and gases as byproducts. In wet combustion the sludge is processed at high temperature (around 550°F) and under pressure. This results in the production of ash, gases, and a stabilized liquid.

10.2.6   Disposal

This topic will be covered in detail in Sections 10.4 and 10.5.

10.3   FATE OF VIRUSES DURING SLUDGE PROCESSING

Any pathogen present in wastewater may be partially "transferred" into sludge following wastewater treatment. Thus sludge is a potential health hazard, since it contains high concentrations of viruses, bacterial pathogens, and protozoan and helminth parasites. Protozoan cysts and helminth eggs survive sludge treatment, particularly anaerobic digestion. Although cysts are generally sensitive to drying, helminth ova, particularly those of *Ascaris lumbricoides*, may survive for extended periods of time in sludge and soil.

In recent years, the virus content of sludge has been investigated and concentrations ranging from 5000 to 28,000 PFU/l of raw sludge have been reported. Technology for virus detection in sludge is still plagued with problems, as follows.

Cytotoxicity. Sludge may be toxic to host cells when sludge eluates are assayed directly on tissue cultures. The toxicity due to heavy metals has been successfully handled by extracting sludge eluates with dithizone in chloroform. Dialysis may also help in the detoxification of sludge eluates.

Virus Dissociation from Sludge. The majority of viruses in sludge are solid-associated. Therefore, any detection method must take into account these solid-bound viruses. Sludge treatment with beef extract, fetal calf serum or sodium dodecyl sulfate (at pH ranging from 7.5 to 9.0) followed by mechanical shaking (vortexing or sonication) generally releases some of the bound viruses. The efficiency of these eluents was obtained from artificially contaminated sludge. Elution with beef extract followed by organic flocculation has resulted in approximately 30% virus recovery. More than 50% of added viruses (poliovirus 1, echovirus 7, coxsackievirus B3) are recovered through treatment of sludge with 0.05 M glycine at pH 11.5. The high pH may, however, be detrimental to some viruses.

Sludge is considered safe when the most sensitive detection method does not result in any virus isolation.

We have so far considered the problems encountered when one attempts to recover viruses from sludge. We will now examine the fate of viruses following sludge processing.

10.3.1 Anaerobic Digestion of Sludge

Of all the sludge treatment processes, anaerobic digestion has received most of the attention. This treatment does not completely destroy viruses. Enteric viruses survive in sludge which has been subjected to mesophilic (30–32°C) or thermophilic digestion (50°C) for many days or even weeks. For example, coxsackievirus B5 can survive thermophilic digestion for a period of 30 days.

Laboratory experiments have been undertaken to study the inactivation of viruses during anaerobic digestion. Data obtained to date show that many enteroviruses (echovirus 11, poliovirus 1, coxsackievirus B4, coxsackievirus A9) and coliphage MS2 survive a 48-hour digestion period at 35°C (Figure 10.3). In other seeded-sludge experiments poliovirus was recovered after 30 days of

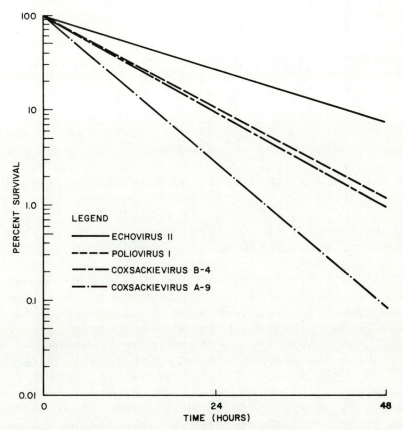

FIG. 10.3. Enterovirus inactivation during anaerobic digestion of sludge. Adapted from J. Bertucci et al. (1975), *48th Ann. Water Poll. Contr. Fed. Conf., Miami Beach, Fla.*

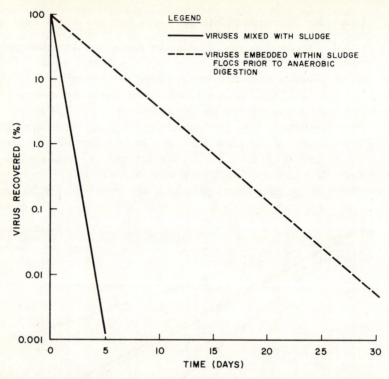

FIG. 10.4. Poliovirus inactivation during anaerobic digestion of sludge. From B. E. Moore et al. (1976), in: *3rd Nat. Conf.: Sludge Management, Disposal, and Utilization, Miami, Fla.* Courtesy of Information Transfer, Inc.

anaerobic digestion (Figure 10.4). This longer survival time may be more realistic, since viruses were embedded within sludge flocs before being subjected to anaerobic digestion.

At Sandia Laboratories (Albuquerque, New Mexico) research was undertaken to understand better the mechanisms of virus inactivation in digested sludge. It was first learned that poliovirus is significantly more inactivated in digested sludge than in raw sludge. Later it was discovered that the virucidal agent in digested sludge is the *uncharged form of ammonia*. It is well known that the concentration of the uncharged form of ammonia (NH_3) is pH-dependent. Table 10.2 shows that as the pH increases from 7 to 10, the concentration of NH_4^+ (charged form of ammonia) decreases from above 99% to 13%. Thus, an increase of the pH of the sludge from 7 to 10 should bring about a higher virus inactivation. Figure 10.5 demonstrates this phenomenon for poliovirus suspended in NH_4Cl, and in digested and raw sludge. The virucidal effect of ammonia is valid for other enteric viruses. Table 10.3 shows that reovirus is most resistant to ammonia. This virus may therefore be suitable as an indicator of the effect of anaerobic digestion on viruses.

It is possible that other virucidal agents are involved in the inactivation of viruses during sludge digestion. Ammonia does not alter significantly the viral capsid, and the RNA core appears to be the main target. Little is known about the extent of viral inactivation during aerobic digestion of sludge. Sludge monitoring at Jay, Florida, revealed that aerobically digested sludge contains more

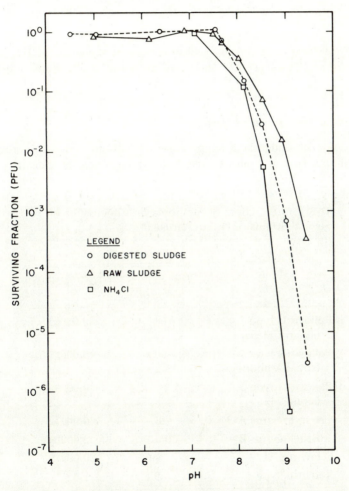

FIG. 10.5. Comparative virucidal (poliovirus 1) effects of NH$_4$Cl, digested sludge and raw sludge as a function of pH. From R. L. Ward and C. S. Ashley (1977), *Appl. Environ. Microbiol.* **33**:860–864. Courtesy of American Society of Microbiology.

TABLE 10.2.   Variation of NH_4^+
Percentage with pH

pH	% NH_4^+
7	>99
8	95
9	59
10	13

virus (426 $TCID_{50}/g$) than anaerobically digested sludge (8 $TCID_{50}/g$). It is suspected that viruses are not effectively destroyed after aerobic digestion of sludge.

10.3.2   Heat Treatment of Sludge

Pasteurization of sludge at a temperature of 70°C for 1 hour destroys effectively helminth eggs and most bacterial and viral pathogens. This treatment is

TABLE 10.3.   Effect of Ammonia on Infectivities of
Several Strains of Enteric Viruses[a]

Virus	Recovery (%) of PFU	
	Tris, pH 9.5 (Control)	NH_4Cl, pH 9.5
Poliovirus type 1 (strain CHAT)	63	<0.000035
Poliovirus type 1 (strain Mahoney)	100	<0.000014
Poliovirus type 2 (strain 712)	60	<0.000044
Coxsackievirus A13	28	<0.00012
Coxsackievirus B1	100	<0.000046
Echovirus 11	9.6	<0.00015
Reovirus type 3	9.8	3.1

[a] Adapted from R. L. Ward and C. S. Ashley (1977), *Appl. Environ. Microbiol.* **33**:860–864.

nonetheless costly and energy-intensive. Moderate heat treatment was proposed following intensive research on the mechanisms of heat inactivation of viruses in sludges. Raw sludge displays a protective effect toward poliovirus that has been exposed to temperatures of 43 and 51°C (Figure 10.6). The protective agent is associated with the sludge solids and is concentrated during sludge drying. It follows that heat inactivation of enteroviruses is much slower in dry (80% solids) than in liquid sludge (5% solids) (Figure 10.7). An investigation on the nature of this protective agent has revealed that it is a heat-stable ionic detergent. This substance has been isolated, and its protective effect has been demonstrated and furthermore shown to be decreased by ammonia, the virucidal agent associated with the liquid portion of digested sludge (Table 10.4). Thus it is possible to counteract the protective effect of the agent by increasing the pH of the sludge to 8.4–9.0. Detergents are, however, responsible for increasing the rate of heat inactivation of reoviruses. This virucidal action toward reoviruses increases with pH. Therefore, at least under laboratory conditions, the solution to the problem of heat inactivation of viruses in sludge seems straightforward: *A moderate heat treatment under alkaline conditions may be effective in inactivation of both enteroviruses and reoviruses.*

Figure 10.8 summarizes the mechanisms involved in heat inactivation of viruses in sludge.

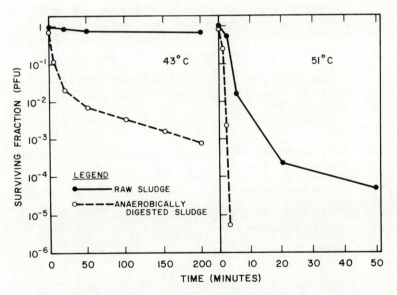

FIG. 10.6. Effect of sludge on the rate of inactivation of poliovirus. Adapted from R. L. Ward and C. S. Ashley (1976), *Appl. Environ. Microbiol.* **32**:339–346.

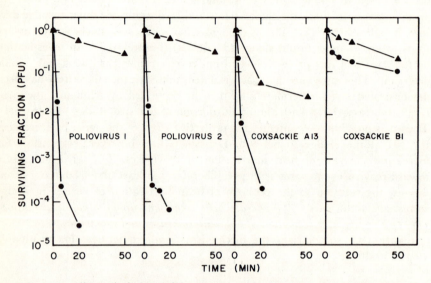

LEGEND
● 5 % SOLIDS
▲ 80 % SOLIDS

FIG. 10.7. Effect of sludge solid content on heat inactivation (50°C) of four enteroviruses in raw sludge. Adapted from R. L. Ward and C. S. Ashley (1978), *Appl. Environ. Microbiol.* **36:**889–897.

TABLE 10.4. Effect of Sludge Solids-associated Agent (Ionic Detergent) on the Heat Activation of Poliovirus 1[a]

Sample	Poliovirus Recovery (PFU/ml)
Unheated control	3.1×10^8
PBS (phosphate buffered saline)	8.5×10^4
Sludge agent (pH 7)	2.8×10^8
Sludge agent (pH 9)	1.8×10^6
Sludge agent + 50 mM NH$_4$Cl (pH 7)	1.7×10^8
Sludge agent + 50 mM NH$_4$Cl (pH 9)	5.0×10^2

[a] From R. L. Ward and C. S. Ashley (1977), *Appl. Environ. Microbiol.* **34:**681–688. Courtesy of Amer. Soc. Microbiol.

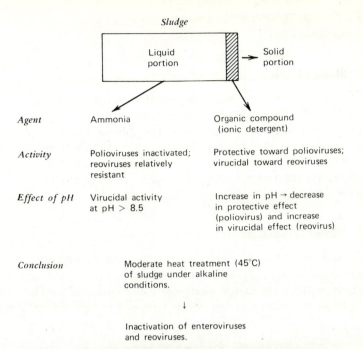

FIG. 10.8.   Mechanisms involved in the heat inactivation of viruses in sludge. Adapted from R. L. Ward and C. S. Ashley (1977), *Appl. Environ. Microbiol.* **34**:681–688; and from Ward and Ashely (1978), *Appl. Environ. Microbiol.* **36**:889–897.

10.3.3   Sludge Drying

Spreading of sludge on drying beds raises its solid content from 5% to approximately 40% within 10 to 15 days. Examination of dried sludge in Florida has revealed the presence of enteroviruses (echoviruses) in samples from drying beds. Research in Canada has shown that 41% of lagoon-dried sludge samples were positive for viruses. These alarming data have prompted some research on the effect of drying on viruses in controlled laboratory experiments.

The inactivation of enteric viruses (poliovirus, coxsackievirus, and reovirus) in sludge dried for 11 days is described in Table 10.5. Virus decline is low when the sludge solids content is raised from 5 to 30 or 40%. However, there is a 4 to 5 $\log_{10}$ decrease in virus numbers when the final sludge solid precentage is above 90. Dewatering causes the release of viral RNA and inactivation appears to be due to the dewatering process itself and not to some virucidal agent.

These findings show that air-drying of sludge in drying beds does not completely inactivate viruses.

10.3.4   Sludge Liming

Phosphate removal from sewage is sometimes achieved by addition of lime (calcium hydroxide). The high pH (=11.5) resulting from this treatment, apart from removing phosphate, brings about a significant reduction of pathogenic bacteria and viruses. The survival of poliovirus in the lime-sludge produced has been studied and the results are shown in Table 10.6. The virus is undetectable after 12 hours in the sludge. Liming of sludge may thus be an efficient method for destroying viruses.

10.3.5   Sludge Irradiation

In radiation processing of sludge, one makes use of high-energy electrons produced by an electron accelerator or radiation sources such as cesium-137. A pilot plant, using high energy electrons, was constructed in Deer Island in Boston, Mass. A diagram of the system is shown in Figure 10.9. An electron

TABLE 10.5.   Recovery of Some Enteric Viruses After 11 Days of Dewatering of Sludge by Evaporation[a]

Final % of Sludge Solids	Recovery (PFU/ml)
Poliovirus (CHAT)	
5	6.5×10^6
30	3.8×10^6
91	$<2.5 \times 10^2$
Coxsackievirus B1	
5	1.3×10^7
33	3.0×10^6
93	$<2.0 \times 10^2$
Reovirus	
5	1.2×10^6
40	7.0×10^5
94	$<2.0 \times 10^2$

[a] Adapted from R. L. Ward and C. S. Ashley (1977), ASM Conference, New Orleans, La.

TABLE 10.6.   Inactivation of Poliovirus in Lime-Sludge[a]

Run	Total PFU in Sludge Stored at 28°C for:		
	0 hour	6 hours	12 hours
1	4.6×10^3	5.0×10^2	N.D.[b]
2	0.6×10^3	2.0×10^2	N.D.
3	0.2×10^3	4.5×10^2	N.D.
4	0.1×10^3	0.3×10^2	N.D.

[a] Adapted from S. A. Sattar, S. Ramia, and J. C. N. West-wood (1976), *Can. J. Publ. Health* **67**:221–225.

[b] N.D. = not detected.

accelerator tube irradiates a thin layer (about 2 mm) of sludge spread on a rotating drum. The sludge is passed at a velocity of 2 m/s and the irradiation dose is 400 krads. The radiation region is protected by a 6-ft-thick concrete box. This system was designed to treat about 100,000 gal of sludge per day. It has been argued that this method is of low cost as compared with other sludge treatments.

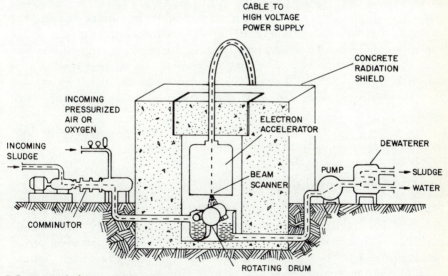

FIG. 10.9.   Sludge irradiation: Diagram of Deer Island, Mass., electron treatment system. From Trump (1975), *in: Williamsburg Conf. on Management of Wastewater Residuals*, J. L. Smith and E. H. Bryan, Eds.

Irradiation of dry and liquid sludge with cesium-137 is also practiced at Sandia Laboratories, Alburquerque, New Mexico.

Microbial cells are generally highly susceptible to ionizing radiation. The radiosensitive target is the nucleic acid. With respect to radioresistance, there is an inverse relationship between microbial size and mean lethal dose. Ionizing radiations inactivate microorganisms either directly or indirectly via production of free radicals.

Pathogenic bacteria are efficiently reduced (6–8 log) at radiation doses ranging from 200 to 300 krad (1 kilorad = 1000 rad). The regrowth of coliforms and pathogenic bacteria (e.g., *Salmonella*) in irradiated sludge (200 krad) has been demonstrated. This problem is, however, not encountered in heat pasteurized sludge or sludge irradiated at or above 400 Krad.

Helminth eggs are also effectively destroyed by radiations and a 3 log_{10} (99.9%) reduction is obtained with a radiation dose of 90 krad at room temperature.

Viruses appear to be more resistant to ionizing radiations than pathogenic bacteria or parasites. A radiation dose of 1 Mrad (1 Megarad = 1000 krad) is needed to bring about a 2–3 log_{10} reduction of poliovirus 2. Other viruses are somewhat more susceptible and their radioresistance is displayed in Figure 10.10.

Thermoradiation is a process that combines ionizing radiation with heat, resulting in a synergetic effect on microbial pathogens and parasites. This process also improves some physical characteristics of the sludge, namely, solids settling and filterability. Thermoradiation readily inactivates viruses. For example, a radiation dose of 150–200 krad, combined with heating at 51°C, can reduce poliovirus 1 numbers by more than five orders of magnitude in only 5 minutes. The sensitivity of various microorganisms to thermoradiation is shown in Table 10.7. Fecal streptococci appear to be the most resistant and may eventually serve as indicators for virus presence in thermoirradiated sludge.

Irradiation of dry or composted sludge results in a product that can be used as an animal feed supplement or for agricultural purposes.

10.3.6   Sludge Composting

Sludge composting is a thermophilic (55–70°C) and aerobic process that results in a stabilized and odor-free product. Sludge is mixed with other organic materials such as woodchips or leaves and allowed to decompose for a period ranging from 3 weeks in summer to more than one month in winter. Aerobic conditions are maintained either by pumping air into the compost pile or by regularly turning the pile. Two methods of sludge composting are used in the United States, the *windrow* and the *pile system*.

TABLE 10.7. Comparative Sensitivity of
Viruses and Bacteria to
Thermoradiation[a]

Organism	KRADS[b]	Min at 60°C[b]
Adenovirus	450	0.15
Poliovirus	300	1.5
Coliforms	20	2.0
Salmonella	45	7.5
Streptococci	120–200	15.0

[a] Adapted from J. R. Brandon (1976), *Water Sew. Wks.* Sept. 1976.

[b] Radiation dose and heating time required for 1 Log inactivation.

1. ***Windrow System.*** Dewatered sludge is mixed with dried sludge and placed in a *windrow* (i.e., row) for a period ranging from 21 (summer) to 40 days (winter). The windrow is approximately 10 feet wide and 4 feet deep. Air is provided by turning the sludge on a daily basis. This system is used in Los Angeles, California Figure 10.11 shows a picture of the windrow system. While the windrow system appears suitable for digested sludge, it is unacceptable for raw sludge due to odor problems.

2. ***Aerated Pile System.*** (e.g., Beltsville Process.) This process has been developed at the Agricultural Research Center, Beltsville, Md. Raw or digested sludge is mixed with woodchips and placed over a woodchip base. The mixture is then covered with screened compost and allowed to decompose for a period of 3 weeks. Aerobic conditions are maintained by drawing air into the compost pile. The pile is then spread and allowed to dry. The compost is marketed after screening to recycle the woodchips and after a 30-day storage (i.e., curing) period. A flow chart of the process is shown in Figure 10.12.

The two composting processes produce an odor-free, stabilized material that has some value as a fertilizer and as a conditioner that benefits the physical properties of soils. This material improves soil structure, aeration, and water holding capacity. However, compost is not necessarily pathogen-free and some further treatment (e.g., irradiation) may be necessary, especially in cases when this product is used for growing edible crops.

Composting, being a thermophilic process with temperatures ranging from 60 to 70°C under ideal conditions generally results in the inactivation of patho-

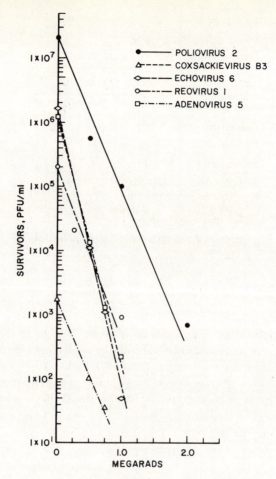

FIG. 10.10 Radioresistance of enteric viruses in sewage treatment plant effluents. From T. G. Metcalf (1975), *in: Williamsburg Conf. on Management of Wastewater Residuals*. J. L. Smith and E. H. Bryan, Eds.

genic microorganisms and parasites. Temperatures above 55°C are maintained for several days and appear to be responsible for the destruction of helminth ova. Figure 10.13 illustrates the extent of destruction of *Ascaris* eggs in composted sludge. Temperatures of 55°C or above readily reduce the embryonation of these parasites. Protozoan cysts follow similar trends during composting. It is thus expected that a properly operating composting system should efficiently eliminate parasite eggs and protozoan cysts.

FIG. 10.11.   Sludge composting: Windrow system. Courtesy of W. D. Burge.

Pathogenic bacteria are also effectively inactivated during the digestion and storage (curing) phases. Their die-off increases with depth. Some may, however, multiply at the surface of the compost pile.

Experiments with model viruses (bacterial phage f2) have revealed that a properly operating windrow system may result in total virus inactivation in approximately 50 days (Figure 10.14). Enteric virus monitoring of a compost system has revealed their presence during the windrow phase but none after the curing period.

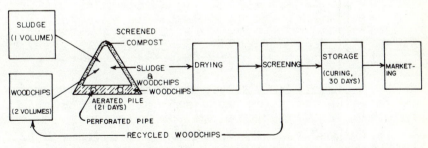

FIG. 10.12.   Sludge composting: Beltsville aerated pile method. Adapted from E. Epstein (1977), *Compost Sci.* **18:**5–7; and from Parr et al. (1978), *Agric. Environ.* **4:**123–137.

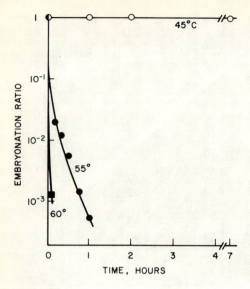

FIG. 10.13. Heat inactivation of Ascaris eggs in composted sludge (60% solids). Adapted from J. R. Brandon (1978), *Parasites in Soil*/Sludge Systems. SAND 77-1970 Report. Sandia Lab., Albuquerque, N.M.

We have shown in Section 10.3.2 that the rates of heat inactivation of enteric viruses in sludge are greatly influenced by ionic detergents. While detergents protect enteroviruses from thermal inactivation, they are responsible for increased decline of reoviruses. These chemicals are substantially degraded during composting and this results in different rates of inactivation in composted

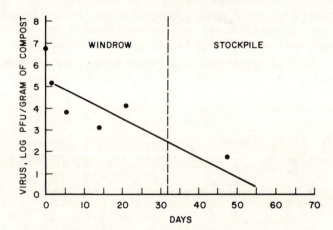

FIG. 10.14. Inactivation of bacterial phage f2 during composting of digested sludge. From K. Kawata et al. (1977), *Water Sew. Wks.* **124**:76–79. Courtesy of Water Sewage Works.

LEGEND

■ RAW SLUDGE, POLIOVIRUS I

▲ COMPOSTED SLUDGE, POLIOVIRUS I

□ RAW SLUDGE, REOVIRUS

△ COMPOSTED SLUDGE, REOVIRUS

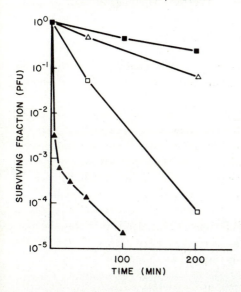

FIG. 10.15. Comparative rates of heat inactivation (45°C) of poliovirus and reovirus in raw and composted sludge. Adapted from R. L. Ward and C. S. Ashley. 1978. Appl. Environ. Microbiol. **36:**898–905.

sludge as compared to raw sludge. Figure 10.15 shows that poliovirus is more thermally stable in raw than in composted sludge, while reovirus displays a reverse pattern. This means that the high temperatures (55–70°C) generated during composting will inactivate efficiently most of the enteroviruses. However, reoviruses may survive during this thermophilic process.

Compost is a useful product that may be of some interest to public (e.g., landscaping of public buildings, cemeteries, reclamation of marginal lands) and private (e.g., nurseries, golf courses) agencies. The safety aspect of its use for edible crops by residential gardeners is still unclear.

10.4   SLUDGE DISPOSAL

Sludge disposal is undoubtedly one of the most difficult problems facing communities around the globe. In the United States federal and state regulations require higher treatment levels than in the past for municipal wastewaters. This has accentuated the problem because more sludge will be produced around the country. The law requires acceptable methods for the utilization and disposal of sludge. Over the years various methods have been used for the disposal of this cumbersome material. These methods include incineration, landfill, ocean dumping, and land application. Sludge is disposed of according to the percentages shown in Table 10.8.

Incineration. Incineration requires the burning of sludge at high temperatures. It produces ashes that must be disposed of in landfills. This practice is plagued by problems of air pollution and high fuel costs.

Landfill. Sludge is buried in conventional sanitary landfills. This alternative may result in groundwater and surface water pollution.

Ocean Discharge. This method is obviously used only by coastal cities. The digested sludge is carried offshore on barges (New York Metropolitan Area) or pumped into deep water (Boston). In 1977, 7.4 million tons of sludge were dumped at sea, most of it in the Atlantic Ocean. This practice will be phased out by the end of 1981 because of concern over public health (sludge may be transported back by currents to public beaches) and damage to marine life.

Land Application of Sludge. Land disposal of sludge is probably the most promising method in the future. It has been proposed to use sludge for the reclamation of strip-mined areas. The soil pH is raised and enough organic

TABLE 10.8.   Estimate of Current Sludge Ultimate Disposal and Utilization Practices[a]

Method	Treated Sludge 1000's dry T/yr	Percentage of Total
Ocean disposal	529	9.1%
Lagoons[b]	264	4.5%
Land application (incl. giveaway/sales)	1366	23.4%
Landfill	1277	21.9%
Incineration	(2400)[c]	41.1%
Total	5836	100.0%

[a] Courtesy of R. K. Bastian, U.S. EPA, Office of Water Prog. Oper. (1979)

[b] Lagoons generally (but not always) are only a temporary sludge disposal mechanism and the sludge placed in lagoons will usually be disposed of by one of the other means eventually.

[c] Estimated amount of raw sludge (dry tons) currently incinerated.

matter is provided to allow the reestablishment of vegetation in those areas. Sludge disposal on agricultural land seems to be the most attractive and economically viable solution to the sludge problem.

Presently, less than 0.3% of U.S. croplands receive sludge, and it is estimated that only 1% of the croplands would be sufficient to receive all the sludge produced in the United States.

10.5   LAND DISPOSAL OF SLUDGE

Agricultural soils are being seriously considered for the disposal of sewage effluents and residuals. The subject of disposal of sewage effluents has been examined in Chapter 9. Application of sludge onto soil may have some beneficial and some detrimental effects.

10.5.1   Beneficial Aspects

The following benefits have been considered.

1. The *nutrients* present in the sludge are advantageously used by crops.
2. *Improvement of soil properties.* The structure and the water-retention capacity of the soil are improved following the addition of a material with high organic matter content.

10.5.2   Detrimental Aspects

Sludge application to land may result in the following.

Excess Salt Concentration. Sludge contains high salt concentrations which may be toxic to some plants. Plants may also be affected by gases evolving from sludge and by low oxygen.

Nitrate Pollution of Groundwater. Nitrogen present in the sludge may, when in nitrate form, be transported easily through the soil profile and contaminate the groundwater supplies. High nitrate concentrations have been associated with the occurrence of *methemoglobinemia,* a disease striking infants under the age of 4 months ("blue-babies"). Methemoglobinemia is characterized by the inability of red blood cells to carry oxygen and may lead to asphyxia.

Toxic Organic Chemicals. The compounds of most concern are organochlorine insecticides (e.g., aldrin, dieldrin, endrin, kepone) and chlorinated phenolics.

Heavy Metal Toxicity. Sludge contains a variety of heavy metals such as cadmium (Cd), copper (Cu), chromium (Cr), zinc (Zn), lead (Pb), nickel (Ni), and mercury (Hg). Some of these trace metals are not essential to plant growth (Hg, Cd, Cr, Ni). Therefore, they may be incorporated into the food chain after sludge application to land and thereafter threaten public health. Of most concern is cadmium, which is readily absorbed, translocated, and accumulated by certain crops. Most municipal sludges contain less than 20 ppm of cadmium. Since food is the main source of cadmium for humans, concern has been raised over the potential accumulation of cadmium in food crops and animal meat. This heavy metal may accumulate in kidneys and cause renal troubles.

The U.S. Environmental Protection Agency (EPA) has established guidelines regulating the application of sludge to land according to the heavy metal status of the sludge. EPA specifies that the cadmium should not exceed 10 mg/kg dry sludge and that the Zn : Cd ratio should be equal to or more than 100. This ratio has been based on the findings that high Zn levels reduce Cd uptake by plants. These guidelines are shown in Table 10.9. As a comparison, the levels of some heavy metals in digested sludge from the city of Pensacola, Florida are also included. The high Zn level of this particular sludge proved to be toxic to certain plants.

Some feel that the EPA guidelines are too strict and that the metal levels are not related to the total amount of sludge applied to land.

TABLE 10.9.   EPA Guidelines: Maximum Content of Heavy Metals for Sludges to be Applied to Agricultural Lands[a]

Element	Pensacola (Fla.) liquid digested sludge (mg/kg dry sludge)	EPA guidelines: Maximum level (mg/kg dry sludge)
Cd	7	10
Cu	309	1000
Hg	5	10
Ni	20	200
Pb	193	1000
Zn	2570	2000

[a] Adapted from J. Bertrand et al. (1978), *1st Ann. Conf. Appl. Res. and Pract. Mun. Ind. Waste, Madison, Wis.;* and USEPA Document; "Policy Statement on Acceptable Methods for the Utilization or Disposal of Sludges"; EPA, Washington, D.C. 1974.

The heavy metal problem is more severe in acidic than in alkaline soils. In the latter heavy metals revert to forms that are unavailable for plant uptake. Therefore in sludge-treated acidic soils the addition of lime would be of some benefit. Obviously, the prevention of entry of heavy metals into industrial wastes would be a better approach.

Public Objections. The public may object to the land disposal of sludge because of odor and aesthetic (attraction of flies) problems. These may be solved by injecting sludge into the soil.

Problem of Viral Pathogens. The virus content of sludge was examined and it was shown that many typical sludge treatment operations do not completely remove the virus threat to humans. Virus-laden sludge may potentially contaminate surface water and groundwater when applied to land.

The potential routes of virus transmission following sludge disposal are shown in Figure 10.16.

10.5.3 Fate of Viruses During Sludge Application to Land

Sludge may be applied to land mainly via the following methods.

1. *Spray Irrigation.* This practice may create an aerosol problem, which is discussed in Chapter 9.
2. *Surface Spreading.* Sludge is spread on the soil surface and mixed with the soil within 2 to 10 days. According to conditions prevailing on the land application site, this practice may lead to contamination of natural waters via surface runoff. This can be minimized by plowing the topsoil–sludge mixture as soon as possible.
3. *Subsurface Injection.* Sludge is injected at 10 to 20 cm below the soil surface at rates up to 3000 liters/minute. This practice avoids problems due to aerosols, surface runoff, pests, and odors, and minimizes human exposure to sludge.

Whatever the land disposal scheme adopted, one must be concerned with the transport pattern and survival of sludge-associated viruses within the soil matrix.

Virus survival is controlled by many environmental factors, the most important being temperature and soil moisture. In Denmark, at temperatures ranging approximately from 0 to 10°C (monthly averages), it took 161 days to reach a 5 $\log_{10}$ reduction of coxsackievirus B3 in sludge-amended soil. In Florida, at temperatures ranging between 21 and 33°C, it took only 21 days to reduce poliovirus 1 numbers by more than 5 orders of magnitude (Table 10.10). Virus was detected after 21 days under the hot and wet conditions of

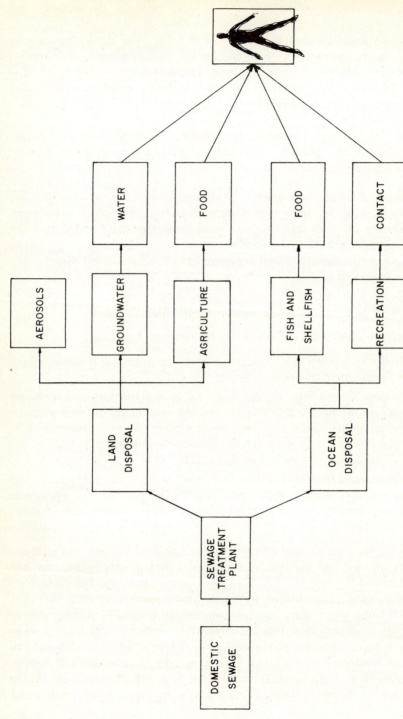

FIG. 10.16. Potential routes of virus transmission via sludge disposal. Adapted from D. E. Weaver et al. (1977), *in: Land as a Waste Management Alternative.* R. C. Loehr, Ed., Ann Arbor Sci., Ann Arbor, Mich.

TABLE 10.10.   Survival of Poliovirus Type 1 Following Sludge Application to a Sandy Soil (Temperature range: 21.4–32.6°C)[a]

Date	Number of Days	Rainfall (cm) (Cumulative)	Soil Moisture %	Solar Radiation (MJ/m2) (Cumulative)	Number of Viruses PFU/g Soil or Sludge Core 1	Core 2
Sludge Sampling						
06/02/78	0	0		16.4	5.5×10^7	5.5×10^7
06/03/78	1	0.50		28.6	1.5×10^7	1.6×10^6

At Day 4 Sludge was mixed with top 2.5 cm of soil

Date	Number of Days	Rainfall (cm) (Cumulative)	Soil Moisture %	Solar Radiation (MJ/m2) (Cumulative)	Number of Viruses PFU/g Soil or Sludge Core 1	Core 2
Soil Sampling (top 2.5 cm)						
06/06/78	4	6.12	17.24	75.9	3.2×10^4	4.4×10^4
06/07/78	5	6.12	0.84	98.5	8.6×10^4	7.5×10^3
06/09/78	7	7.39	11.48	137.3	2.5×10^3	2.4×10^3
06/23/78	21	10.61	0.16	424.5	4.6	15.1
07/07/78	35	13.84	0.21	721.9	1.3	0.9

[a] From G. Bitton, O. C. Pancorbo, S. R. Farrah, J. M. Davidson, and A. R. Overman (unpublished). Undisturbed columns (40 cm length and 15.5 cm diameter) of Eustis Fine Sand received 2.5 cm of lagoon sludge (3% solids) seeded with poliovirus type 1. The total number of PFU applied to soil was 7.8×10^8. Virus monitoring in soil was undertaken according to the isoelectric casein method. All soil leachates were collected and monitored for viruses, pH, and conductivity. The soil columns were exposed to natural conditions during the summer of 1978 in Gainesville, Florida.

the Floridian summer but none (poliovirus 1 and echovirus 1) was recovered under hot and dry conditions in the fall (Table 10.11). Low temperatures tend to prolong virus survival in sludge-amended soils. Poliovirus 1 was able to persist more than 3 months under winter conditions in Cincinnati, Ohio. Moreover, in Butte, Montana, indigenous viruses were shown to survive the winter freeze in a sludge-injected soil. Subsurface injection of sludge, despite its aesthetic acceptability and other advantages, may allow a longer survival of viruses as compared with surface spreading.

The topic of virus transport through soils has been reviewed in Chapter 9. Viruses are retained effectively by the sludge–soil matrix and their transport to the groundwater is thus minimized. Research undertaken at the University of Florida has revealed that soil cores incubated under natural conditions and

TABLE 10.11. Survival of Poliovirus 1 and Echovirus 1 Following Sludge Application to a Sandy Soil (Temperature range: 15.0–28.3°C)[a]

Date	Number of Days	Rainfall in cm (Cumulative)	Soil Moisture (%)	Solar Radiation (MJ/m2) (Cumulative)	Number of Viruses PFU/g Sludge or Soil			
					Poliovirus 1		Echovirus 1	
					Core 1	Core 2	Core 1	Core 2
Sludge Sampling								
10/11/78	0	0	—	12.2	2.7×10^7	2.7×10^7	9.0×10^4	9.0×10^4
10/14/78	3	0	—	45.1	1.3×10^6	2.9×10^6	1.6×10^4	1.1×10^4
At Day 3 Sludge was Mixed with Top 2.5 cm of Soil								
Soil Sampling (top 2.5 cm)								
10/14/78	3	0	8.14	45.1	4.3×10^4	1.9×10^4	2.8×10^2	1.9×10^2
10/16/78	5	0	3.11	83.9	3.5×10^3	2.1×10^2	6.9×10^1	1.6×10^1
10/19/78	8	0	1.01	121.7	3.6×10^3	2.1×10^2	5.6×10^1	6.3×10^1
11/01/79	21	0	1.01	300.2	N.D.[b]	N.D.	N.D.	N.D.

[a] From G. Bitton, O. C. Pancorbo, S. R. Farrah, J. M. Davidson, and A. R. Overman (unpublished). Undisturbed columns (40 cm length, 15.5 cm diameter) of Eustis fine sand received 2.5 cm of lagoon sludge (7% solids) seeded with poliovirus 1 or echovirus 1. The total number of virus applied to soil was 8.8×10^8 PFU for poliovirus 1 and 2.9×10^6 PFU for echovirus 1. Virus monitoring in soil was undertaken according to the isoelectric casein method. All soil leachates were collected and monitored for viruses, pH, and conductivity. The soil columns were exposed to natural conditions during the fall of 1978 in Gainesville, Fla.

[b] Not detected.

treated with 2.5 cm of seeded sludge remove more than 99.999% of poliovirus type 1 after 30 cm of rain. In Denmark under cold conditions soils ranging from 5 to 21% clay were able to completely retain sludge-associated coxsackievirus B3.

A field monitoring program was undertaken to follow the fate of viruses in a sludge-application site in Jay, Florida. No virus was detected in groundwater samples (190 to 380 liters) during a one-year monitoring program.

10.6   SUMMARY

Viruses are transferred to sludge following their passage through sewage treatment plants.

Viruses are not completely destroyed by anaerobic digestion. The virucidal agent in digested sludge is the uncharged form of ammonia.

Sludge pasteurization destroys virus effectively. Investigation of the mechanism(s) of thermal inactivation of viruses in sludge has revealed that it is possible to destroy enteroviruses and reovirus by heat treatment under alkaline conditions.

Air-drying of sludge in drying beds does not completely inactivate viruses.

Thermoradiation, a process that combines radiation with heat, readily inactivates viruses.

Sludge composting, a thermophilic and aerobic process, results in the destruction of parasites and microbial pathogens, including most of the enteroviruses. Reoviruses may survive this thermophilic process.

Land application of sludge is an attractive alternative to other disposal methods. Heavy metal toxicity and viral pollution are of most concern regarding sludge application onto land. Virus survival in sludge-amended soils is greatly dependent upon temperature and soil moisture. In regard to their transport pattern through soils, viruses are effectively retained by the sludge–soil matrix.

10.7   FURTHER READING

Cliver, D. O. 1976. Surface application of municipal sludge. In: *Virus Aspects of Applying Municipal Waste to Land,* L. B. Baldwin, J. M. Davidson, and J. F. Gerber, Eds. University of Florida, Gainesville, Fla.

Fair, G. M., J. C. Geyer, and D. A. Okun, 1968. *Water and Wastewater Engineering,* Vol. 2: *Water Purification and Wastewater Treatment and Disposal,* Wiley, New York.

Loehr, R. D., Ed. 1977. *Land as a Waste Management Alternative.* Ann Arbor Sci., Pub., Ann Arbor, Mich.

Lund, E. 1974. Disposal of sludge. In: *Viruses in Water.* G. Berg, et al., Eds. American Public Health Association, Washington, D.C.

Metcalf and Eddy, Inc. 1972. *Wastewater Engineering,* McGraw-Hill, New York.

Moore, B. E., B. P. Sagik, and C. A. Sorber, 1976. An assessment of potential health risks associated with land disposal of residual sludge. In: 3rd National Conf.: *Sludge Management, Disposal and Utilization,* Miami, Fla.

Sagik, B. P., and C. A. Sorber, Eds. 1978. *Risk Assessment and Health Effects of Land Application of Municipal Wastewater and Sludges.* Center for Applied Research and Technology, The University of Texas, San Antonio, Texas.

Smith, J. L., and E. H. Bryan, Eds., 1975. *Williamsburg Conference on Management of Wastewater Residuals.* Williamsburg, Va.

Sopper, W. E., and L. T. Kardos, Eds., 1973. *Recycling Treated Municipal Wastewater and Sludge Through Forest and Cropland.* Pennsylvania State University Press, University Park, Pa.

eleven
Viruses in Human Food

11.1 INTRODUCTION

Foods have been known to serve as a vehicle for the transmission of numerous diseases afflicting man. These diseases may be due to the following:

- *Bacteria.* They may be the cause of salmonellosis or brucellosis. Bacteria such as *Clostridium botulinum* are also responsible for food poisoning such as botulism.
- *Protozoa.* Toxoplasmosis is caused by *Toxoplasma gondii.*
- *Nematodes.* Trichinosis is contracted through consumption of pork infected with *Trichinella spiralis.*
- *Cestodes.* Parasitic flatworms (beef, pork, and fish tapeworms).
- *Viruses.* Foods may serve as a vehicle for the transmission of at least one viral disease, infectious hepatitis. This disease has characteristic

symptoms that render it suitable for study by epidemiologists. Other viruses, such as the Newcastle disease virus, may cause eye infection in poultry plant workers. The possibility exists that enteric viruses, other than infectious hepatitis virus, may be transmitted via contaminated food. It is clear that epidemiologists have paid little attention to foodborne viruses because such studies are expensive and rather complicated. It is, for example, difficult to distinguish between viral infections due to direct contact and those due to food. Therefore the significance of viral contamination of food remains to be properly assessed.

As a result of the development of adequate detection methods, viruses, especially those of human origin, have been found in many types of foods including oysters, clams, mussels, fish, milk, eggs, vegetables, and ground beef. These findings are of particular significance, since some of these foodstuffs are eaten raw (shellfish, fish, eggs), thus precluding any process that may cause total or at least partial inactivation of viruses.

Foods usually provide a good milieu for the persistence of viruses, but it is unlikely that virus multiplication takes place in any food. However, viruses present a potential for human infection, even though they are present in small numbers in food. We will now examine the ways by which foodstuffs may become contaminated with viruses.

- *Primary contamination.* Food may become contaminated prior to the time of slaughter. Some human viruses have been found in animals which then may act as a vehicle for the transmission of viral diseases. However, viruses of animal origin are probably not a real threat to human health.

- *Secondary contamination.* This is the contamination that occurs after the time of slaughter or harvesting. The main source of secondary contamination is via food handling during preparation at home and in restaurants or during food processing. Contamination may also occur through polluted wastewater effluents, polluted surfaces, flies, cockroaches, and rodents. We will put some emphasis on the use of wastewater effluents for food crop production in agriculture and seafood production in aquaculture systems.

It is probable that infection of consumers by contaminated food could lead to further infections as a result of contact transmission within a susceptible population.

11.2   VIRUSES IN MEAT AND MILK

Although some animal viruses may cause disease in man (foot and mouth disease, Newcastle disease), their impact on human health is largely unknown.

TABLE 11.1.   Viral Isolation from Ground-Beef Samples
Purchased from Markets[a]

Source	Number of Meat Loaves Examined	Average PFU/5 g	Virus Identified
9 different markets	9	0	—
3 different markets	1	195	Poliovirus 1 and Echovirus 6
	1	6	Poliovirus 2
	1	1	Poliovirus 1 and Poliovirus 3

[a] From R. Sullivan, A. C. Fassolitis, and R. B. Read, Jr. (1970), *J. Food Protection* **35**:624. Courtesy of Internat. Assoc. Milk, Food, Environ. Sanitarians, Inc., Ames, Iowa.

Furthermore, some human viruses have been demonstrated in domestic animals such as swine, cows, and goats. Meat may therefore be contaminated at the source (primary contamination) or by infected workers during the numerous handling operations (secondary contamination). Beef is heavily consumed in the United States (over 10 million tons of beef consumed in 1973) as ground meat

TABLE 11.2.   Raw Milk-Associated Viral Diseases[a]

Number of Cases	Probable Source of Virus
Poliomyelitis	
6	Child living on dairy farm (via flies)
8	Infected milker
62	Unknown, but all from one dealer
≧18	Unknown, but flies were common in milk
Infectious Hepatitis	
14	Contaminated water used to wash utensils
3	Incubating individual milked cows and handled milk until onset of jaundice

[a] Adapted from D. O. Cliver (1967), *Health Lab. Sci.* **4**:213–221.

(hamburger), and virus monitoring of market samples has revealed the presence of enteroviruses in concentrations ranging from 1 to 195 PFU/5 g of meat (Table 11.1). The viruses identified were poliovirus 1, 2, and 3, and echovirus 6. This is of particular significance to those consuming raw meat (e.g., "steak tartar") but would not be a health hazard when the meat is well cooked.

Raw milk has been associated with outbreaks of poliomyelitis and infectious hepatitis since around 1900 (Table 11.2). Milk is usually contaminated by infected food handlers, polluted utensils, and flies. These flies have been shown to act as vectors for viruses and to be capable of disseminating these pathogens within a community. Viruses may survive for weeks in flies fed with feces from poliomyelitis patients. Other insects, such as cockroaches, may also serve as potential vectors for enteroviruses.

The advent of pasteurization has decreased, if not eradicated, disease outbreaks associated with milk consumption. However, secondary contamination is a distinct possibility following improper handling of pasteurized milk.

11.3   VIRUSES IN VEGETABLES AND FRUITS

Vegetables are subject to contamination by food handlers after harvesting. It has been reported that between 1967 and 1969 contaminated vegetables and fruits were responsible for approximately 9% of the foodborne outbreaks in the United States (Figure 11.1). Prior to harvesting vegetables may also be contaminated by wastewater effluents used for irrigation purposes. This practice has been responsible for many disease outbreaks caused by bacteria, protozoa, parasitic helminths, and viruses (Table 11.3). Viral hepatitis has been caused by consumption of watercress contaminated with septic tank effluents. Apart from this particular instance, no documented outbreak of viral disease has been attributed to the consumption of vegetable crops following contamination with wastewater or improperly treated wastewater effluents.

The threat of vegetableborne viral diseases has triggered some worldwide interest in the fate of viruses on vegetables irrigated with sludge or sewage effluents.

A serious threat probably stems from vegetables which have a relatively short growth period (e.g., radishes) and are eaten raw. Table 11.4 illustrates the persistence of poliovirus on vegetables (lettuce and radish) spray-irrigated with sewage sludge and effluents in the area of Cincinnati, Ohio. The virus was able to survive from 14 to 36 days under field conditions. These data are significant, since a 36-day period is enough for planting and harvesting crops such as radishes. Temperature and solar radiation are probably the most important environmental factors that control virus survival on the surface (especially leaf surface) of vegetables. Thus, there is a potential hazard of viral disease follow-

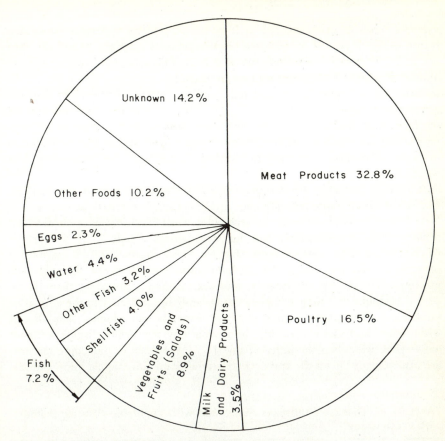

FIG. 11.1.   Vehicles responsible for reported foodborne outbreaks in the United States, 1967–1969. From F. L. Bryan (1973), in: *Microbial Safety of Fishery Products*, C. O. Chichester and H. D. Graham, Eds. Academic Press, New York. Courtesy of Academic Press.

ing consumption of raw vegetables grown in soils that have been spray-irrigated or flooded with sewage sludge and effluents. We have dealt so far with the deposition of viruses on vegetable surfaces and their survival under field conditions. Other pertinent questions have been raised concerning the absorption of viruses by plant roots and their translocation to the aerial parts (stems and leaves) of the plant. Experimentation on potato and tomato roots immersed in a poliovirus suspension did not show any translocation of the viruses to stems or leaves. However, mouse encephalomyelitis virus was sometimes translocated to the aerial parts of pea, potato, or tomato plants. It is quite conceivable that viruses may enter into plants through root lesions. We still lack information on crops that are indirectly used for human consumption. For example, corn is primarily fed to livestock or undergoes further processing before being consumed by humans. Once vegetables enter the household, they are generally stored in a refrigerator at 4–6°C. This relatively low temperature allows the survival of enteric viruses on vegetable surfaces under moist conditions for up to 2 months. In addition, routine washing of vegetables does not completely remove viruses that may be present on the surface, and this depends on the type of vegetable under consideration. In contrast to tomatoes and peppers which have smooth surfaces, lettuce has a rugged surface that is difficult to clean and that allows the prevalence of moist conditions, thus prolonging virus survival in a refrigerator.

It appears that fruits are less implicated in disease outbreaks than vegetables. Fruits have been suspected only once as the cause of an infectious outbreak in New Jersey, in 1965. Contrary to vegetables, some fruits display some antiviral activity. This is particularly true for strawberries (Table 11.5) and grapes (Table 11.6). The agents responsible for virus inactivation in grapes are phenolic compounds which are predominantly located in the skin. Phenols inactivate viruses by temporarily complexing them. However, there is no reason to doubt that these complexes may be dissociated once in the animal or human gut. The phenolic content of red wines is much higher than that of white wines and this is probably the reason why the former display a higher antiviral activity than the latter (Table 11.7). Polyphenols, including tannins, are also responsible for the antiviral activity of grape juice, apple juice, and tea.

TABLE 11.3.   Disease Outbreaks Associated with Vegetables Contaminated with Wastewater[a]

Disease	Source of Contamination	Vegetable
Typhoid fever	Sewage-contaminated watercress bed	Watercress
Typhoid fever	Sewage irrigation	Raw vegetables
Salmonellosis	Sewage irrigation and sludge	Vegetables
Salmonellosis	Primary treated effluent	Cabbage
Amebiasis	Sewage irrigation	Vegetables
Ascariasis	Sewage spray irrigation	Vegetables
Hookworm infection	Sewage farming	Vegetables
Fascioliasis	Sewage polluted water	Watercress
Viral hepatitis	Septic tank effluent	Watercress

[a] Adapted from E. L. Bryan (1977), *J. Food Protection* **40**:45–56.

TABLE 11.4.   Persistence of Poliovirus on Vegetables
Spray-Irrigated with Inoculated Sewage Sludge and
Effluent (PFU/g of plant)[a]

	Sludge Irrigated		Effluent Irrigated	
Day Tested	Lettuce	Radish	Lettuce	Radish
A. August–October 1973[b]				
1	172	156	25	154
2	112	294	14	7
4	23	7	3	3
5	14	13	2	1
8	0.8	6	2	0.6
14	1	0.2	0	0.1
17	0.2	0.4	0.1	0.2
27	0.1	0.1	0	0
36	0.1	0.1	0.1	0
B. June–July 1974[c]				
1	180	950	20	172
3	4	300	3	80
6	1.4	46	0.5	10
8	0.2	14	0.4	7
10	0	0.4	0.2	0.2
14	0.2	0.6	0	0

[a] Adapted from E. P. Larkin, J. T. Tierney, and R. Sullivan (1976),
J. Environ. Eng. Div. **102**:29–35.
[b] Temperature was 15–33°C.
[c] Temperature was 19–33°C.

These beverages may thus act as natural antiviral agents, but their efficacy
probably depends on the stability of the complexes they form with viral agents.
These complexes may be destabilized in the human gut, but this remains to be
demonstrated.

   In conclusion, outbreaks of viral diseases may potentially occur after irrigation of vegetable crops with wastewater sludge and effluents. This risk should

TABLE 11.5.   % Recovery of Viruses
from Fruits after Storage for 24 Hours
at 4°C[a,b]

Fruit	Coxsackie B5	Echo 7
Strawberry	4–11	0–6
Cherry	85–99	95–105

[a] Adapted from J. Konowalchuk and J. I. Speirs
(1975), *J. Milk Food Technol.* **38**:598–600.

[b] The fruits were washed with water.

be greater in countries that use raw or poorly treated sewage (e.g., sewage farms in India). In the United States the use of raw sewage or primarily treated effluents for the irrigation of edible crops is prohibited by most of the states. Some states allow irrigation of vegetable crops with only disinfected sewage effluents.

11.4   VIRUS IN SEAFOOD

11.4.1   Anatomy and Ecology of Shellfish

Shellfish are of great importance to the fishery industry because they can be grown in large quantities in relatively small estuarine areas and because they are a relatively expensive delicacy. Unfortunately, they are also important vectors of bacterial, viral, and parasitic diseases.

Shellfish are bivalve mollusks (Phylum: Mollusca; Class: Pelecypoda) that grow in estuarine areas. This group is comprised of oysters, clams, and

TABLE 11.6.   Recovery of Poliovirus
from Grape Extract and Infusion
(pH = 7) after Incubation for 24 Hours
at 4°C[a]

Type of Grape	% of Input Virus	
	Extract	Infusion
Tokay	<1	17
Ribier	<1	21

[a] Adapted from J. Konowalchuk and J. I. Speirs
(1976), *Appl. Environ. Microbiol.* **32**:757–763.

TABLE 11.7.   Virus Survival (%) in Wine After Incubation for 24 Hours at 4°C[a]

Wine Type	Wine pH	Dilution of Wine in Water		
		Undiluted	10^{-1}	10^{-4}
French red	3.8	<1	8	87
Italian red	3.4	<1	<1	63
German white	3.3	9	53	94

[a] Adapted from J. Konowalchuk and J. I. Speirs (1976), *Appl. Environ. Microbiol.* **32**:757–763.

mussels. Table 11.8 lists important edible shellfish species in North America. The shell enclosing the body is composed of two valves, which are hinged together dorsally. Bivalve mollusks feed by pumping large quantities of estuarine water. The volume of pumped water may vary from 4 to more than 20 liters/hour. The pumping rate varies with shellfish species and is affected by environmental factors such as temperature, salinity, and turbidity.

The pumping, due to ciliary action, brings in small particles (phytoplankton, bacteria, viruses), which are trapped on the mucus secreted on the gills. Then, the mucus is pushed by ciliary action toward the mouth, and the

TABLE 11.8.   Some North American Edible Shellfish Species[a]

Scientific Name	Common Name
Crassostrea virginica	Eastern oyster
Crassostrea gigas	Pacific oyster
Mya arenaria	Soft-shell clam
Mercenaria campechiensis	Southern quahaug
Mytilus edulis	Blue mussel
Mercenaria mercenaria	Hard-shell clam, Northern quahaug
Ostrea lurida	Olympia oyster
Saxidomus giganteus	Butter clam

[a] From T. G. Metcalf (1976), *in: Virus Aspects of Applying Municipal Wastes to Land*, L. B. Baldwin et al., Eds., University of Florida, Gainesville, Fla.

microorganisms enter the stomach. Within the stomach, microbial particles may be digested by extracellular enzymes or may be transported to the digestive diverticulum. They eventually reach the intestine and are later shed in feces. Shellfish feed selectively on suspended particles, some of which are rejected and incorporated into *pseudofeces*. As to their habitat, oysters and mussels are found attached to surfaces, whereas clams are buried into the sand and pump water through siphons (Figure 11.2).

11.4.2   Shellfish-associated Diseases and Hazards

There are three main features that determine the importance of shellfish as vectors of human diseases.

- *Ecology of Shellfish.* These animals live in estuarine environments, which are most subject to pollution by human wastes.
- *Feeding Habits of Shellfish.* They concentrate viruses and other microorganisms by pumping large quantities of estuarine water.
- *Preparation of Shellfish.* They are often eaten raw or insufficiently cooked.

Thus, fishery products in general, and shellfish in particular, are the source of many diseases of bacterial, viral, protozoan, and helminthic origin. Among the bacterial diseases cholera, typhoid fever, and gastroenteritis have been traced to the consumption of raw or insufficiently cooked shellfish. *Vibrio parahaemolyticus* is particularly important, since it is responsible for 50 to 70% of foodborne diseases in Japan. This bacterium is most abundant in summer and it causes gastroenteritis.

As compared to bacterial diseases, the only viral diseases known to be associated with shellfish are infectious hepatitis and gastroenteritis. The

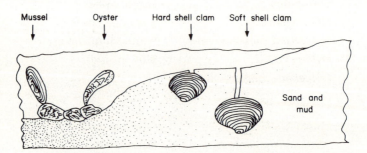

FIG. 11.2.   Estuarine habitat of shellfish. From T. G. Metcalf (1976), in: *Virus Aspects of Applying Municipal Wastes to Land,* L. B. Baldwin et al., Eds. University of Florida, Gainesville.

TABLE 11.9.   Some Shellfish-Associated Infectious Hepatitis Outbreaks[a]

Number of Cases	Shellfish	Source of Shellfish
629	Oysters	Held in polluted by awaiting sale
84	Oysters	Mouth of Pascagoula River, Mississippi
263	Oysters	Houston, Texas (oysters from Louisiana)
481	Clams	Raritan Bay, New Jersey
258	Clams	Narragansett Bay, Rhode Island
7	Mussels	Melbourne, Australia

[a] Adapted from D. O. Cliver (1964), *Health Lab. Sci.* **4:**213–221 and from J. L. Dienstag et al. (1976), *Lancet* **1:**561–563.

association of infectious hepatitis with shellfish consumption was first reported in Sweden in 1955 and later confirmed in the United States in 1961. Most of the reported epidemics have been since linked to uncooked shellfish. The epidemiology of shellfish-associated infectious hepatitis is relatively simple as compared to other viral diseases because the outbreaks are recognized by distinctive clinical symptoms (see Chapter 2). Some shellfish-associated outbreaks are shown in Table 11.9. A particular outbreak is the one reported in Australia where 7 individuals out of a family of 16 developed infectious hepatitis one month after a family picnic. These individuals consumed incompletely cooked mussels (*Mytilus edulis*). The disease was comfirmed serologically (detection of hepatitis A antigen in stools) and a relationship was shown between shellfish consumption and development of the disease (Table 11.10).

The Norwalk agent was recently implicated in a large outbreak (more than 2000 cases) of oyster-associated gastroenteritis in Australia.

TABLE 11.10.   Relationship of Mussel Ingestion to the Development of Infectious Hepatitis (Melbourne, Australia)[a]

	Developed Hepatitis		
Ate Mussels	Yes	No	Attack Rate (%)
Yes	7	3	70
No	0	4	0

[a] Adapted from J. L. Dienstag et al. (1976), *Lancet* **1:**561–563.

TABLE 11.11. Some Examples of Virus Isolation from Shellfish[a]

Species of Shellfish	Source	Virus Type Isolated	% Positive
C. virginica	Little and Great Bay, New Hampshire, known to be polluted with raw sewage	Echo 9, Coxsackie B4	6 of 10 pools of 10 oysters each
Mussels	Italian markets of Bari and Parma	Echo 3, 9 & 13	50 samples of 3 mussels each
C. virginica	New Hampshire estuaries (for 4 years)	Polio 1, 2, 3; echo 9; reo 1; coxsackie B2, B3, B4	114 of 459 pools
Mya. arenaria	Area known to be bacteriologically polluted	Hepatitis type B antigen	1 of 4
Mytilus galloprovincialis	Leghorn coast of Italy	Echo 5, 6, 8, 12; coxsackie A18	5 of 68 pools of 10 mussels each
C. virginica	Estuarine inlet of North Atlantic Ocean	1 enterovirus, 30 coliphages	59 of 130 pools positive either for coliphage or enterovirus
(i) oysters (ii) mussels	Poitiers, France	Coxsackie A18 predominant	(i) 7 of 70 pools; (ii) 2 of 10 pools

Many other enteric viruses have been isolated from shellfish on many occasions (Table 11.11), but epidemiological evidence for their transmission by shellfish is not well established.

Other hazards associated with shellfish are due to the tendency of these mollusks to concentrate dinoflagellate toxins, heavy metals, hydrocarbons, pesticides, and radionuclides. Of most concern is the *paralytic shellfish poisoning* (PSP) caused by the concentration of dinoflagellates (*Gymnodinium brevis*, *Gonyaulax catanella*) by shellfish. These algae produce heat-stable toxins that cause paralysis and neurotoxic symptoms in shellfish consumers. One can

TABLE 11.11. (Continued)

Species of Shellfish	Source	Virus Type Isolated	% Positive
C. virginica	Galveston Bay, Texas	Polio 1, 2	28.6%
Mer. mercenaria	Atlantic coast of Maine	Hepatitis type B antigen	100% samples positive from this site but none from 20 other sites
C. virginica	(i) Texas Gulf coast	Echo 1, polio 1	2 of 17 samples
	(ii) Louisiana Gulf Coast	polio 3	1 of 24 samples
	(iii) frozen and shucked oyster imported from Japan	polio 1	1 of 1 sample
Mer. Californiacus	Collected from stations remote from sewage outfalls; placed in cages and suspended at various depths below buoys located offshore near outfall diffusers		18 of 39 samples
Clams and oysters	Great South Bay and Oyster Bay, Long Island, N.Y.	Polio 1, 2 echo 2, 15, 20, 23 coxsackie B3	3 of 22 in open waters; 5 of 24 in closed waters

[a] Adapted from C. P. Gerba and S. M. Goyal (1978), *J. Food Protection* **41**:743–754.

monitor for PSP-associated toxicity by the mouse bioassay. The area is closed for catching shellfish when the toxicity reaches 400 mouse units/100 g shellfish.

11.4.3 Uptake and Elimination of Viruses by Shellfish

Shellfish growing in sewage-contaminated estuaries take up and rapidly accumulate viruses, bacteria, yeasts, algae, and other particles.

Stationary and flow-through experiments have shown that oysters accumulate substantial amounts of viruses within 12 to 48 hours. The uptake rate

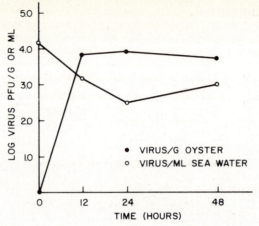

FIG. 11.3. Uptake of poliovirus by *Ostrea lurida* in contaminated seawater for 48 hours (Salinity = 28%; temperature = 13°C). From R. DiGirolamo et al. (1975), *Appl. Microbiol.* **29**:260–264. Courtesy of American Society of Microbiology.

varies from one species to another and is enhanced when viruses are solids-associated. Figure 11.3 illustrates the uptake of poliovirus by the Olympia oyster, *Ostrea lurida*. We have seen that food particles are generally entrapped within a mucus mass secreted by the shellfish. Then the mucus is pushed into the mouth by ciliary action. It has been shown that 38 to 68% of enteroviruses (Table 11.12) are attached to the mucus mass and more specifically to sulfate radicals by ionic bonding. The attachment process is dependent on environ-

TABLE 11.12.   Uptake of Enteroviruses by Shellfish Mucus[a]

Virus Type	Virus Uptake by Shellfish Mucus (%)
Polio 1	68.0
Coxsackie A9	53.0
Coxsackie B4	53.0
Echo 1	38.0

[a] Adapted from R. DiGirolamo et al. (1977), *Appl. Environ. Microbiol.* **33**:19–25.

TABLE 11.13.   Accumulation of Poliovirus by Oysters[a]

Time (hr)	Virus (PFU/g or ml of sample)			
	Seawater	Mantle, Gills, Fluid	Digestive Area and Feces	Body
Crassostrea gigas (Pacific oyster)				
0	1.9×10^4	0	0	0
48	1.2×10^3	3.0×10^3	1.2×10^4	3.0×10^3
Ostrea lurida (Olympia oyster)				
0	2.0×10^4	0	0	0
48	1.2×10^3	3.9×10^3	1.3×10^4	1.3×10^3

[a] Adapted from R. DiGirolamo et al. (1975), *Appl. Microbiol.* **29**:260–264.

mental factors such as salinity and pH. Viruses accumulate mainly in the shellfish digestive area, and many diffuse afterward into the body tissues. This specific accumulation in the digestive area is illustrated in Table 11.13 for two west coast oyster species. This obviously is of public health significance, since the entire animal is consumed raw.

When placed in clean water (sterilization of water may be accomplished by chlorination or ozonation), contaminated shellfish species have the ability to cleanse themselves from viruses, bacteria, yeasts, and silt. This cleansing process is called *depuration*. Oysters depurate themselves within 48 to 72 hours (Figure 11.4). However, some viruses, especially those associated with the body

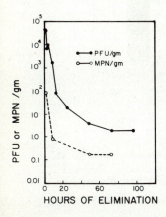

FIG. 11.4. Elimination of poliovirus and *E. coli* by eastern oysters (flow-through system). Adapted from J. R. Mitchell et al. (1966), *Am. J. Pub. Health* **84**:40–50.

tissues, may remain within the oyster for longer periods of time. Moreover, it was found that oysters may retain *Salmonella typhimurium* for up to 14 days. This indicates that commercial depuration based on the removal of fecal coliforms within 48 hours may not be sufficient to obtain seafood that is safe for human consumption. It has been found that under certain conditions coliform indicators are not reliable indicators of the presence of enteric viruses in shellfish and in estuarine water. This subject is examined in more detail in Chapter 12.

Viruses not eliminated by the depuration process may remain within the

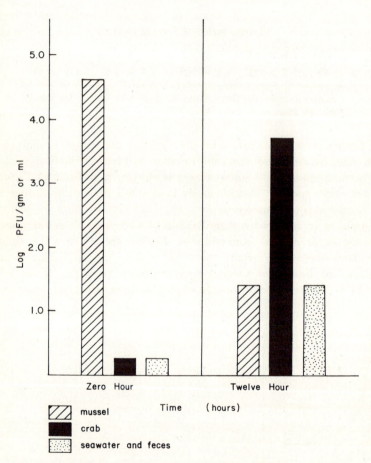

FIG. 11.5. Uptake of poliovirus by shore crabs allowed to feed for 12 hours on contaminated mussels (*Mytilus californiacus*). *Source* R. DiGirolamo et al. (1972), *Appl. Microbiol.* **23:**170–171. Courtesy of American Society of Microbiology.

shellfish tissues for relatively long periods of time. Their survival is temperature-dependent and increases when the temperature of water decreases.
Shellfish do not seem to produce any antibody-like substances that may inactivate viruses.

On a commercial basis, shellfish are treated by the following:

- **Relaying.** Shellfish are transferred to clean areas.
- **Depuration.** Shellfish are held in basins or tanks where seawater of adequate salinity has been treated by ultraviolet or ozone.

It has been recently demonstrated that human-associated yeasts, *Candida
parapsilosis, C. tropicalis,* and *Torulopsis glabrata,* are frequently isolated
from oysters, mussels, and quahogs. The relationship of yeast occurrence in
shellfish to human health is not clear yet.

11.4.4   Transfer of Viruses Through the Food Chain

Another important aspect of shellfish contamination by enteric viruses is the
transfer of these pathogens to higher trophic levels of the food chain. For
example, crabs become contaminated when feeding upon mussels. Figure 11.5
shows the uptake of poliovirus by shore crabs feeding on contaminated mussels
(*Mytilus californiacus*). It is also known that enteric viruses may be transferred from shellfish feces to polychaete worms and then to detritus-feeding
fishes. These hypothetical transmission routes are considered in our examination of aquaculture systems that use nonchlorinated wastewater effluents as a
nutrient source.

11.5   FATE OF VIRUSES IN AQUACULTURE SYSTEMS

In marine aquaculture systems, food, in the form of commercial fertilizers, can
account for an important part of the operating costs. It has thus been proposed
to use wastewater effluents as a source of nutrients in aquaculture. This system
serves two main purposes:

1.   Food production.
2.   Tertiary treatment of wastewater effluents (nutrient removal).

The flow chart of a typical aquaculture system is shown in Figure 11.6.
Seawater–wastewater effluent mixtures enter a phytoplankton (e.g., *Skel-
etonema costatum, Phaeodactylum tricornutum*) tank. This step results in the
removal of soluble nutrients such as nitrate and phosphate. The algae will, in
turn, serve as a food source for shellfish (oysters, clams, mussels). Soluble
nutrients produced by the shellfish help support the growth of commercial

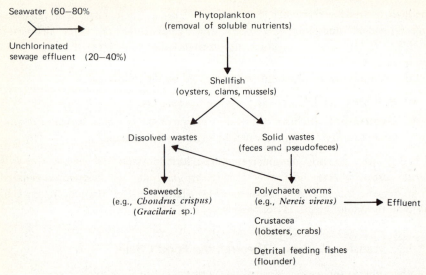

FIG. 11.6. Flow chart showing a typical aquaculture system. Adapted from J. H. Ryther and J. C. Goldman (1975), in: *Water Quality Management through Biological Control*, P. L. Brezonik and J. L. Fox, Eds., University of Florida, Gainesville.

seaweeds (e.g., *Gracilaria* sp., *Chondrus crispus*). Seaweeds are commercially valuable, since they are the source of polysaccharides, *carrageenan* (from *Chondrus crispus*), and *agar* (from *Gracilaria* sp.), which are used as thickening agents in pharmaceutical, cosmetic, and food industries. These seaweeds can bring an income as high as $400–600/dry ton. The solid wastes excreted by the shellfish support the growth of crustacea (crabs, lobsters), detrital fishes (e.g., mullet) or polychaete worms (e.g., *Nereis virens*).

At Woods Hole Oceanographic Institution (Woods Hole, Mass.), a pilot facility has been constructed to process as much as 100,000 gal of wastewater effluent per day. A diagram of the flow-through system at Woods Hole is shown in Figure 11.6.

Despite the advantages of aquaculture systems, concern was raised over the following:

- *Public Health.* Pathogens and chemicals (mainly heavy metals) may be bioconcentrated in the food chain.
- *Climate Variations.* A change in temperature and light intensity results in changes in the growth of algae, the primary producers.
- *Space Requirements.*
- *Nutrient Imbalances in the Wastewater.*

We are obviously concerned here with the fate of viruses in aquaculture systems.

We know that viruses are not completely removed by biological treatment of municipal wastewater. Thus, through the use of wastewater effluents as a nutrient source one can expect pathogenic viruses to enter the aquaculture system. We will now follow these pathogens as they go through the system:

1. *Algal Ponds.* Sunlight and temperature may bring about some viral inactivation. One is interested here in the interaction of virus with algae, the major biological component in these ponds. It has been found that indeed virus survival may be enhanced by the presence of algae and by organic materials in sewage effluents. However, the heavy growth of

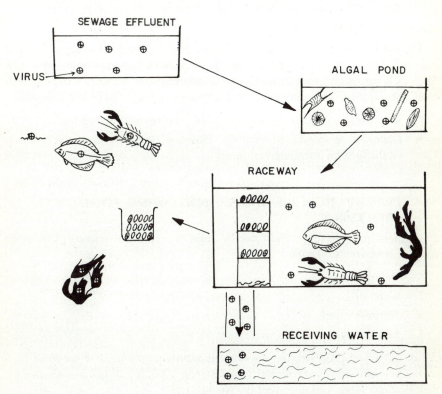

FIG. 11.7.   Virus survival and transmission potential within a mariculture system. From T. G. Metcalf et al. (1977), in: *Risk Assessment and Health Effects of Land Application of Municipal Wastewater and Sludge*, B. P. Sagik and C. A. Sorber, Eds., University of Texas, San Antonio. Courtesy of Center for Applied Research and Technology, University of Texas, San Antonio, Texas.

algae may have a detrimental effect toward virus due to increasing pH (see Chapter 6).

2. *Shellfish Tanks.* Viruses can enter the shellfish tanks and be concentrated by the mollusks as described in the previous section. The consumption of the contaminated shellfish is of great concern to public health workers.

3. *Seaweed Tanks.* One should not be too seriously concerned with the contamination of seaweeds, since they are essentially used by the colloid industry. However, virus adsorbed to seaweed may be transmitted to browsing invertebrates.

4. *Detritus-Feeding Organisms.* Crustacea (crabs) feeding on mollusks or their feces may become contaminated and thus present a health hazard. Viruses have been isolated from polychaete worms, feeding on shellfish feces.

Thus a potential transmission of viruses through the food chain is in sight and this threat should be dealt with if one desires the production of a safe food. A schematic representation of virus transmission through a hypothetical food chain is given in Figure 11.7. It is therefore necessary to do the following:

- Remove viruses prior to their entry into the aquaculture system.
- Depurate contaminated shellfish. However, this process is not yet well established on a commercial basis.

11.6   INACTIVATION OF VIRUSES FOLLOWING FOOD PROCESSING

Four main processes are traditionally used to inactivate microorganisms in foods. These are as follows:

1. Heat treatment.
2. Irradiation.
3. Freeze-drying.
4. Addition of chemicals.

All of these food processes were developed mainly on the basis of bacterial control. Their efficiency with regard to pathogenic virus is still being evaluated by food virology laboratories around the world.

11.6.1   Heat Treatment

Heat treatment (cooking) has long been used by man to destroy undesirable microorganisms in foods. The sensitivity of viruses to thermal inactivation varies with the type of food. Moreover, in a particular food viruses may be inactivated at the surface of the food item but persist within the food. This is the reason why enteroviruses were found to persist in hamburgers cooked to "rare" condition. The temperature inside the hamburger, 60°C, is not high enough to kill all the viruses. A similar phenomenon occurs in shellfish subjected to some type of heat treatment. Viruses persist after steaming, frying, and stewing of oysters. In fact steamed clams have been implicated in the transmission of infectious hepatitis. Many factors can modify the outcome of heat treatment in regard to viral inactivation. These factors are pH, cations, proteins, and fats. The first three factors are examined in Chapter 3. Fats have a protective effect toward viruses exposed to heat in milk or in ground meat. Figure 11.8 illustrates this protective effect in meat samples heated to 80°C and containing 3 or 47% of fat.

Milk traditionally undergoes pasteurization prior to human consumption. This process implies the destruction of disease-producing microorganisms. The U.S. Public Health Service recommends that this process be carried out at 62.8°C for 30 minutes or 71.7°C for 15 seconds. Milk can thus be considered reasonably safe following pasteurization. In the United States, egg white is commercially treated by heat-hydrogen peroxide (H_2O_2) pasteurization. The addition of H_2O_2 enables the use of lower temperatures (51.7–54.4°C for 3.5 minutes) to avoid protein denaturation. This milder treatment does not adequately destroy enteroviruses in egg white.

Storage of goods at refrigerator (5–6°C) or freezer (−18°C) temperatures

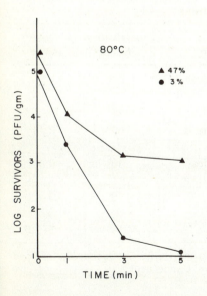

FIG. 11.8. Effect of fat content on thermal inactivation of poliovirus in meat at 80°C. Adapted from J. A. Filippi and G. J. Banwart (1974), *J. Food Sci.* **39**:865.

does not affect virus survival in foods and may allow their persistence for up to 5 months at −20°C.

11.6.2   Food Irradiation

Gamma rays are the predominantly used ionizing radiation in food processing. These are electromagnetic radiations of very short wavelength emitted from radioisotopes such as cobalt-60 (^{60}Co) and cesium-137 (^{137}Cs). The radiation dose is expressed in kilojoules/kg (kJ/kg). One kJ/kg is equal to 10^{-1} Mrad (Mrad = 10^6 rad). Through food irradiation one can reach one of the following objectives:

- **Radappertization.** It allows the destruction of all microorganisms.
- **Radurization.** It brings about the reduction of spoilage micro-organisms.
- **Radicidation.** It aims at inactivating parasites and non-spore-forming pathogens.

Of the three treatments radappertization is the only one involving the use of heat. Heat exposure (73–77°C) is followed by freezing and irradiation, with doses equal to or lower than 10 kJ/kg. We can then assume that viruses are well inactivated during the heat treatment stage of radappertization. Table 11.14 shows the minimal doses of gamma irradiation necessary for the destruc-

TABLE 11.14.   Approximate Minimal Doses of Gamma Irradiation Necessary for the Destruction of Specific Microorganisms[a]

Microorganism	Medium	Inactivation Factor	Dose (Mrad)
C. botulinum type E	broth, minced lean beef	10^6	1.5
Salmonella	broth	10^6	0.32–0.35
E. coli	broth, minced lean beef	10^6	0.18
Streptococcus faecalis	broth, minced lean beef	10^6	0.38
Poliovirus, encephalitis virus	Brain tissue	10^6	3.5–4.0
Vaccinia virus	Buffer	10^6	1.5–3.0

[a] Adapted from N. W. Desrosier and J. N. Desrosier (1977), *The Technology of Food Preservation*, AVI, Publisher Co., Inc., Westport, Connecticut.

TABLE 11.15. Virus Persistence During Freeze-drying of Food[a]

Food	Original Virus Titer (PFU/ml)	Titer After Freeze-drying (PFU/ml)
German potato salad	3.0×10^6	1.5×10^3
Beef and vegetables	3.0×10^6	6.0×10^2
Chicken with gravy	2.6×10^6	2.9×10^2
Salmon salad	2.3×10^6	6.9×10^2

[a] Adapted from N. D. Heidelbaugh and D. J. Giron (1969), *J. Food Sci.* **34**:239.

tion of specific microorganisms. It can be seen that bacteria are generally less resistant than viruses to gamma irradiation. It has been shown that doses of 0.4 to 0.6 Mrads do not completely destroy viruses in fish or shellfish. Irradiation doses higher than 1 Mrad may lead to organoleptic changes in the food.

11.6.3 Freeze-drying of Foods

Freeze-dehydration is also commonly used to extend the shelf-life of various foods. As shown in Table 11.15 this practice cannot be relied on to destroy viruses in foods.

11.6.4 Chemical Inactivation in Foods

We have discussed the presence of natural antiviral chemicals in foods, namely, polyphenols in grapes, grape juice, apple juice, and tea. Food acidity (e.g., orange juice) may be detrimental to some viruses, namely rhinoviruses. The addition of sodium bisulfite ($NaHSO_3$) to cole slaw is also responsible for the inactivation of enteroviruses. It was found that the nucleic acid core is the main target of sodium bisulfite action, although some capsid alteration is also observed.

More information is needed on the effect of chemical food additives on viruses.

11.7 VIRUS DETECTION IN FOOD

In order to assess the significance of foods as vehicles in the transmission of enteric viruses, one has to develop adequate detection methods which must be efficient, accurate, inexpensive, and easy to perform.

Viruses present on fruits and vegetables with a smooth surface are relatively easy to detect. This can be done by rinsing the surface with beef extract, or with glycine buffer at high pH. The eluates may be further concentrated by organic flocculation (beef extract) or by membrane filtration (glycine buffer). However, problems are encountered with other foods which require homogenization and separation of food solids that interfere with viral assay on host cells. These foods (ground meat, vegetables) require an elution step followed by clarification to remove solids. The filtrates are further concentrated by ultrafiltration, ultracentrifugation, or hydroextraction using PEG. The concentrates are then assayed on tissue cultures. Bacterial and fungal contaminants are controlled by antibiotics or by filter sterilization. A flow chart of the detection technique is shown in Figure 11.9. Depending on the food under consideration, these methods allow the recovery of 20 to 100% of viruses present in the food.

Since recent studies suggest a poor correlation between bacterial indicators and virus levels in water and in shellfish, one is compelled to develop adequate techniques for virus detection in shellfish. However, shellfish present a particular problem, that is, their toxicity toward host cell cultures. An adsorption–elution–filtration–concentration technique has been developed at the

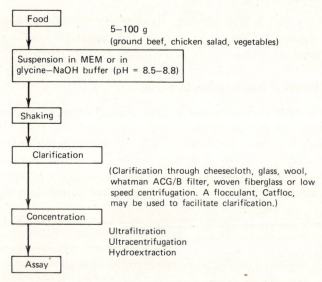

FIG. 11.9. Method for detecting viruses in meats and vegetables. Adapted from Kostenbader and Cliver (1973), *Appl. Microbiol.* **26**:149–154; R. Sullivan et al. (1970), *J. Food Sci.* **35**:624; J. T. Tierney et al. (1973), *Appl. Microbiol.* **26**:497–501.

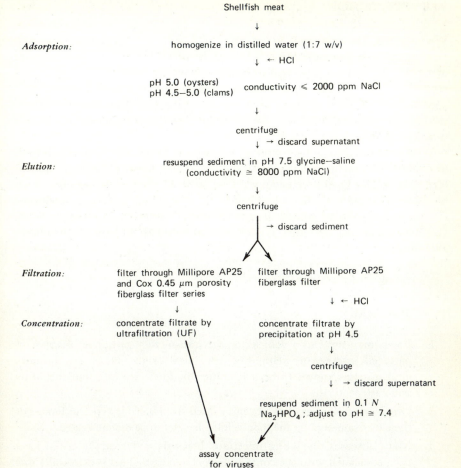

Shellfish meat
↓

Adsorption: homogenize in distilled water (1:7 w/v)
↓ ← HCl

pH 5.0 (oysters)
pH 4.5–5.0 (clams) conductivity ⩽ 2000 ppm NaCl

↓

centrifuge
↓ → discard supernatant

Elution: resuspend sediment in pH 7.5 glycine–saline
(conductivity ≅ 8000 ppm NaCl)

↓

centrifuge

→ discard sediment

Filtration: filter through Millipore AP25 filter through Millipore AP25
and Cox 0.45 μm porosity fiberglass filter
fiberglass filter series
↓ ↓ ← HCl

Concentration: concentrate filtrate by concentrate filtrate by
ultrafiltration (UF) precipitation at pH 4.5
↓

centrifuge
↓ → discard supernatant

resuspend sediment in 0.1 N
Na_2HPO_4 ; adjust to pH ≅ 7.4

assay concentrate
for viruses

FIG. 11.10.   Scheme for recovering viruses for oysters and clams. From M. D. Sobsey et al. (1978), *Appl. Environ. Microbiol.* **36:**121–128. Courtesy of American Society of Microbiology.

University of North Carolina for the recovery of viruses from oysters and clams. This method, outlined in Figure 11.10, utilizes one of two concentration methods, ultrafiltration or acid precipitation. The recovery of viruses from oysters is around 46% for poliovirus 1, reovirus 3, or a simian adenovirus. Since ultrafiltration is too long to perform, acid precipitation would be the preferred method for virus concentration from membrane filter eluates.

11.8   PREVENTION

We have seen that there are many opportunities for viruses to enter food and potentially cause disease in man. This can be prevented by "keeping feces out of food." In addition, appropriate action should be taken by the numerous regulatory agencies that deal with food safety in the United States. These agencies are the Federal Drug Administration, the U.S. Department of Agriculture, the National Marine Fisheries Service, and the Environmental Protection Agency.

Worldwide interest in food virology has led the World Health Organization to establish a committee that compiles all the data related to viruses in human food. The WHO Food Virology program consists of two research groups which have been charged with the task of collecting information concerning foodborne viruses. One group is located at the Veterinary Research Institute in Brno, Czechoslovakia and the other is at the Food Research Institute in Madison, Wisconsin.

11.9   SUMMARY

1. Foods may serve as a vehicle for the transmission of human diseases of bacterial, protozoan, helminthic, and viral origin. However, the only viral diseases known to be transmitted by foods are infectious hepatitis and gastroenteritis.

2. Other enteric viruses have been isolated from various types of food, but their transmission via the food vehicle has yet to be demonstrated.

3. Many diseases, including infectious hepatitis and gastroenteritis, have been associated with the consumption of shellfish. This is essentially due to the ecology of shellfish (they live in polluted estuarine environments), their feeding habits (they pump large quantities of estuarine water), and their preparation (they are eaten raw or insufficiently cooked).

4. The use of improperly treated sewage effluents in marine aquaculture systems may result in the transfer of viruses from the water to the food.

5. In food processing heat treatment is undoubtedly the most reliable method of inactivating viruses.

6. Methodology for virus detection in food has improved during the last few years, and it is now possible to recover up to 50% of viruses from shellfish.

11.10 FURTHER READING

Berg, G. 1964. The food vehicle in virus transmission. *Health Lab. Sci.* **1**:51.

Berg, G. Ed. 1978. *Indicators of Virus in Water and Food.* Ann Arbor Sci., Ann Arbor, Mich.

Bryan, F. L. 1977. Diseases transmitted by foods contaminated by wastewater. *J. Food Protection* **40**:45–56.

Cliver, D. O. 1971. Transmission of viruses through foods. *Crit. Rev. Environ. Control* **1**:551–579.

Gerba, C. P, and S. M. Goyal. 1978. Detection and occurrence of enteric viruses in shellfish: A review. *J. Food Protection* **41**:743–754.

Jay, J. M. 1970. *Modern Food Microbiology.* Van Nostrand Reinhold, New York.

Larkin, E. P. 1978. Foods as vehicles for the transmission of viral diseases. In: *Indicators of Virus in Water and Food,* G. Berg, Ed., Ann Arbor Sci., Ann Arbor, Mich.

Metcalf, T. G. 1976. Prospects for virus infection in man and animals from domestic waste land disposal practice. In: *Virus Aspects of Applying Municipal Waste to Land.* L. B. Baldwin et al., Eds. University of Florida, Gainesville, Fla.

Metcalf, T. G., R. Comeau, R. Mooney, and J. H. Ryther. 1978. Opportunities for virus transport within aquatic and terrestrial environments. In: *Risk Assessment and Health Effects of Land Application of Municipal Wastewater and Sludge.* B. P. Sagik and C. A. Sorber, Eds. University of Texas, San Antonio, Texas.

Rao, V. C. 1976. Virus transmission through foods. *J. Food Sci. Technol.* **13**:287–293.

Stewart, G. T., and M. A. Amerine. 1973. *Introduction to Food Science and Technology.* Academic, New York.

twelve
Indicators of Viruses in Water, Wastewater, and Food

12.1 INTRODUCTION

We have described in Chapter 1 the enteric virus group, and we have shown the wide range of diseases that may be caused by these infectious agents, ranging from skin rash to paralysis. It appears now that classical epidemiologic tools are not sensitive enough to demonstrate the transmission of these viral diseases by the water route. It is possible that waterborne, foodborne, and possible airborne transmission of enteric virus escapes detection, using epidemiologic procedures. It seems thus necessary to develop concentration and rapid assay procedures in order to detect viruses in water and other environmental samples. This topic is covered in some detail in Chapter 5. It seems that this approach is delicate, time-consuming, and rather expensive. However, the presence of viruses is not prime evidence of a public health hazard unless we can correlate it with infection/disease within a community.

Another approach is to look for the "best" indicator or indicators for viruses. Obviously, the "ideal" indicator is the virus itself. The "best" indicator for viruses should have the following characteristics:

1. Should be associated with the source of the pathogen and should be absent in unpolluted areas.

2. Should occur in greater numbers than the pathogen.

3. Should not multiply in water and in other environments.

4. Should be at least equally resistant to environmental stresses and to disinfection as the pathogen.

5. Should be detectable by means of easy, rapid, and inexpensive methods.

6. Should not be pathogenic.

Once a suitable indicator is found, the next step is to find a correlation between the levels of the indicator with disease occurrence by means of prospective epidemiological studies.

At the present time the available indicators of fecal pollution have some limitations:

1. The indicators available are often less resistant than viruses to environmental stresses. They are, on the other hand, sometimes too resistant (e.g., bacterial spores) as compared to viruses, and may thus tend to overestimate viral pollution.

2. The ratio indicator: virus varies. This variation may be due to multiplication of bacterial indicators or to variations in virus numbers.

3. The presence of traditional fecal pollution indicators does signal the possible presence of viruses. Their absence is not, however, indicative of viral absence.

It is useful to review the most important bacterial and other indicators of fecal pollution and examine their relationship to enteric virus presence in environmental samples.

12.2 A SURVEY OF INDICATORS OF FECAL POLLUTION

12.2.1 Coliform Group

Total Coliforms. The total coliform group comprises all of the aerobic and facultative anaerobic, gram-negative, non-spore-forming, rod-shaped bacteria that ferment lactose with gas production within 48 hours at 35°C.

Coliforms are discharged in high numbers in feces, and it has been estimated that each person excretes approximately 2×10^9 coliforms/day. However, some of the members of the coliform group (e.g., *Enterobacter aerogenes*) are not specific to fecal material and may be found in soils and on vegetation. Some coliforms, for example, *Klebsiella* sp., may multiply in pulp and paper mill effluents. Coliforms may also regrow in reservoirs following chlorination of sewage effluents.

Fecal Coliforms. Fecal coliforms comprise all those coliforms that can ferment lactose at 44.5°C. This group reflects more accurately the presence of fecal material from warm-blooded animals. Fecal coliforms are less subject to regrowth in polluted waters than the non-fecal coliforms.

Correlations have been found between waterborne bacterial pathogens such as *Salmonella* and fecal coliforms densities. However, the relationship between fecal coliforms and enteric viruses is not clear, and it will be discussed further in this chapter.

12.2.2 Fecal Streptococci

Fecal streptococci, namely *Streptococcus faecalis,* (25% of the total streptococci) are found in feces of warm-blooded animals including humans. They do not seem to multiply in the aquatic environment and appear to survive better than total or fecal coliforms. The fecal coliform:fecal streptococci (FC:FS) ratio is useful in indicating the source of pollution. (FC/FS) < 0.7 is indicative of animal pollution, whereas (FC/FS) > 4 is indicative of human sources. However, this ratio is only valid for recent (24 hours) fecal pollution. *S. faecalis* survives well in all types of waters, whereas coliforms may or may not survive, depending on water quality.

This group seems to reflect more closely the persistence of viruses in seawater and in sludge digestion than total and fecal coliforms. Nonetheless, the little information available does not allow us to assess their usefulness as indicators for enteric viruses.

12.2.3 Anaerobic Bacterial Indicators

Two major anaerobic bacterial indicators are the clostridia, specifically *Clostridium perfringens,* and bifidobacteria, namely, *Bifidobacterium adolescentis.*

Cl. perfringens is an anaerobic gram-positive, spore-forming, rod-shaped bacterium. Since there are no data comparing its densities in water with those of viruses, one can only speculate on the merits of this bacterium as an indicator of viral pollution. Bacterial spores are very resistant to environmental stresses and to disinfection. They may be, however, too resistant to be useful as indicators for bacteria and even viruses. Since *Cl. perfringens* forms spores that can survive and accumulate in sediments, it has been suggested as an indicator of past pollution and as a tracer to follow the fate of pathogens.

Bifidobacteria are anaerobic, gram-positive, non-spore-forming bacteria. There are 11 species of *Bifidobacterium,* five of which (*B. bifidum, B. adolescentis, B. infantis, B. longum, B. breve*) are primarily associated with humans. Their densities in feces and sewage may be higher than those of coliforms or fecal streptococci. They are less likely to regrow in the extraenteral environment than *E. coli,* and they may be useful for differentiating human from animal fecal pollution. Major disadvantages are their sensitivty to oxygen and lack of adequate methodology for their detection in environmental samples. Their relationship to the enteric virus group has not been documented.

12.2.4 Bacterial Phages

Bacterial phages were discovered by D'Herelle in human feces at the beginning of this century. They occur in large numbers (mean concentration = 500,000 PFU/100 ml) in municipal wastewaters, although some contend that, at least for phages which can be assayed on *E. coli B* or *E. coli K12,* they are infrequently recovered from feces. The total coliform : coliphage ratios vary from 360 in raw wastewater to 10 in treated effluents. This indicates that coliphages are more resistant than coliforms to sewage treatment processes.

The ubiquitous occurrence of coliphages in wastewater in higher numbers than animal viruses and their similar and, sometimes, higher resistance to some water and wastewater treatment processes have led to the suggestion that they may be used as indicators of bacterial and viral pollution. They also offer the advantage of easy, rapid, accurate, and inexpensive assay that may give answers in less than 24 hours. Unfortunately, little is known about their ecology in water and wastewater and particularly about their potential multiplication outside the intestine of warm-blooded animals. However, they can serve as laboratory or field models (i.e., simulants) to assess the removal efficiency of water and wastewater treatment plants, land treatment of wastewater effluents, and disinfection.

A coliphage that has given hope for finding suitable indicator for viruses is f2. It is a single-stranded RNA phage that lyses only the male strain (F^+) of *E. coli* (the host receptor sites are the pili and not the cell wall). This phage closely resembles enteroviruses with regard to many physical and chemical

TABLE 12.1.   Comparison of Some Physical and Chemical Characteristics of Bacteriophage f2 and Poliovirus[a]

Characteristic	f2	Poliovirus
Virion		
Nucleic acid	RNA	RNA
Capsid symmetry	Cubic	Cubic
Envelope	None	None
Capsid diameter (Å)	200–250	270–300
pH stability	3–10	3–10
Nucleic Acid		
No. of strands	1	1
Shape	Linear	Linear
Nucleotides	3300	6000
Protein		
No. of polypeptides	2	4

[a] Adapted from W. N. Cramer et al. (1976), *J. Water Poll. Control Fed.* **48**:61–76.

characteristics (Table 12.1). There is, furthermore, no evidence of its regrowth in wastewater. There is, however, no host bacteria that is specific for f2, and thus little is known about the levels of this phage in feces and in sewage. This bacterial phage is viewed as an acceptable indicator for the enteric virus group, at least under certain circumstances.

This particular phage and other groups of phages (e.g., T series phages which can be assayed on *E. coli B*) can be accepted as valid indicators only after thorough investigation of their relationship with the enteric virus group.

12.3   A DISCUSSION ON THE USEFULNESS OF MICROBIAL INDICATORS IN WATER AND WASTEWATER

12.3.1   Microbial Indicators of the Performance of Water and Wastewater Treatment Plants

Viruses are excreted in variable numbers into municipal wastewaters. They are shed only by infected individuals and primarily by individuals under the age of

15. The extent of their release into the environment also depends on the season of the year, socioeconomic level of the community, and other factors. By contrast, bacterial indicators and possibly coliphages are constantly discharged in large numbers in feces by every segment of the population. It follows that the ratio of indicator bacteria to viruses will vary widely. For example, the fecal coliform: virus ratio varies from $7.5 \times 10^3:1$ to $2.9 \times 10^6:1$ in raw sewage. This ratio displays wide fluctuations following primary and secondary treatment in sewage treatment plants. Therefore, these ratios cannot reflect the virus load in raw sewage, primary and secondary effluents. There are also differences between bacterial indicators and virus with respect to the extent of removal by various unit processes in sewage treatment plants.

Coliphages have been under consideration as indicators of viral pollution in wastewater effluents. Coliphages: virus ratios range within 10^3 to 10^4 in trickling filters and oxidation pond effluents. Researchers at the Technion Israel Institute of Technology have suggested that coliphages can serve as indicators in virus monitoring of wastewaters and other waters.

New coliphages models have also been proposed to evaluate the efficiency of wastewater treatment plants. Researchers in South Africa have proposed *Serratia marcescens* bacteriophage as an indicator of the fate of poliovirus in sewage treatment plants. The phage displayed a higher survival than poliovirus 1 under the conditions studied. Other noncoliphage models, including algal virus LPPl, have been proposed to assess the treatment efficiencies of sewage treatment plants.

With respect to water treatment plants, coliphages have been instrumental in providing important information concerning the performance of water treatment processes such as coagulation, flocculation, sand filtration, or adsorption to activated carbon.

The extent of coliphage presence in water and wastewater treatment plants and their relationship to enteric viruses are still incompletely known.

In future studies researchers should use, for comparative purposes, similar assay systems for coliphages (host cells, incubation conditions) and for enteroviruses (concentration techniques, host cells).

12.3.2   Microbial Indicators of Virus Disinfection

We have reviewed in Section 12.1 the properties of an "ideal" indicator or at least the "best" indicator. With respect to disinfection the indicator should be more resistant than viruses to disinfection.

We will now explore various groups of microorganisms which have been considered as potential indicators for virus disinfection.

Although the total plate count does not differentiate pathogenic from nonpathogenic microorganisms, it was nonetheless proposed for assessing the performance of particular steps (e.g., flocculation, sedimentation) in water treatment plants and the efficiency of wastewater chlorination as well. More recently, in South Africa research was conducted on the most reliable microbial indicators for assessing the efficiency of two wastewater reclamation plants (Stander and Windhoek plants). These plants include lime treatment, storage in an equalization basin, rapid sand filtration, chlorination (2 ppm free chlorine), activated carbon treatment, and final chlorination (2 ppm free chlorine). Water samples were analyzed for total bacteria, total coliforms, enterococci, *Clostridium perfringens, Staphylococcus aureus, Pseudomonas aeruginosa,* coliphages, and enteric viruses. It was found that total bacteria count (total plate count) was the most sensitive indicator of the efficiency of wastewater reclamation plants, particularly during the final disinfection step.

Since coliforms are the classical indicators for water and wastewater disinfection, it would seem reasonable to examine their relationship to viruses. This particular group of bacterial indicators is much less resistant to chlorination than viruses. For example, the resistance of poliovirus 1 to hypochlorous acid may be 1 to 2 orders of magnitude higher than that of *E. coli,* a typical coliform of fecal origin. Due to the increased resistance of viral pathogens, the ratio of coliforms to viruses will considerably decrease following chlorination of sewage effluents. For example, the fecal coliform:virus ratio in primary effluents decreases from 890–720,000:1 to 0.6–34:1 after chlorination. A similar phenomenon is observed in oxidation pond effluents. On the other hand, coliforms may regrow in reservoirs that receive chlorinated sewage effluents and their detection in large numbers may overestimate the threat due to pathogens.

Monitoring of sewage treatment plants has, on many occasions, revealed the presence of viruses in chlorinated sewage effluents. This clearly shows that one cannot rely on the coliform index to assess the viral threat in chlorinated wastewater effluents. The finding of viruses in chlorinated tapwater also reveals the inadequacy of the coliform index.

Anaerobic spore-forming bacteria (e.g., *Clostridium perfringens*) and protozoan cysts have been suggested as possible indicators of viruses. They are, however, too resistant to disinfection and, thus, their presence may have no relationship with virus presence.

Coliphage resistance to chlorination has been investigated and it was observed that some coliphages are more resistant to chlorine than enteroviruses. Table 12.2 compares the percent survival of coliphages and poliovirus type 1 following chlorination of oxidation pond effluents. It appears that coliphages, particularly f2, survive generally better than poliovirus type 1. It was, however, argued that f2 and MS2 are more resistant than enteroviruses following exposure to monochloramine but not to hypochlorous acid.

Other microorganisms may serve as potential indicators of virus inactivation in chlorinated effluents. These indicators include two acid fast bacilli (*Mycobacterium fortuitum* and *Mycobacterium phlei*) and a yeast (*Candida parapsilosis*). These microorganisms survive the chlorination process in wastewater

TABLE 12.2.   Percent Survival of Coliphages and Poliovirus Type 1 Following Chlorination of Oxidation Pond Effluents (Contact Time = 1 Hour)[a]

| | Chlorine Dosage (mg/l) | | | |
| | 20 | | 40 | |
Virus Tested	pH = 6.0	pH = 8.3	pH = 6.0	pH = 8.3
Coliphages (host: *E. coli* B)	13.5	56	1.05	5.8
Coliphages (host: *E. coli* K12)	19.1	52.1	0.5	17.6
f2	46.5	43.5	21.5	64
MS2	21.2	25.5	9.5	22.6
Poliovirus 1	10.2	35.0	14.2	7.0

[a] Adapted from Y. Kott et al. (1974), *Water Res.* **8**:165–171.

treatment plants and are more resistant to chlorine than *E. coli*, *Salmonella typhimurium*, or poliovirus type 1 (Figure 12.1). Furthermore, acid-fast bacilli appear to be more resistant to chlorine than yeasts. The chlorine resistance of these microbes is an advantage in their consideration as valid indicators for viruses following chlorination of water and wastewater.

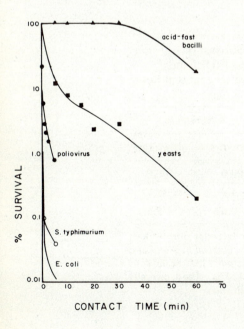

CONTACT   TIME (min)

FIG. 12.1.   Response of pathogens and indicator organisms to 0.5 mg/l free chlorine at pH 7 and 20°C. From R. S. Engelbrecht and E. O. Greening (1978), in: *Indicators of Viruses in Water and Food*, G. Berg, Ed., Ann Arbor Sci. Pub.., Ann Arbor, Mich. Courtesy of Ann Arbor Sci., Pub.

12.3.3 Microbial Indicators in Water, Sediments, and Soils

Virus survival in the aquatic environment is controlled by factors such as temperature, light, pH, and biological factors. It has been well established that temperature is the most significant factor controlling the persistence of viruses outside their host cells. It is well known that viruses persist for long periods in freshwater and marine environments (see Chapters 3 and 4). Their complete disappearance may take days and even weeks as compared to hours in the case of coliform bacteria. In river water viruses have been detected miles away from the source of pollution. In seawater their survival is somewhat shorter, but viruses have been detected as far as 1 mile from sea outfalls. Viruses may also become attached to suspended solids and settle into the sediment where they may remain viable for extended periods of time.

Most of the data accumulated to date point to the longer survival of viruses as compared to bacterial indicators. There have been few reports on correlations between viruses and coliform numbers in the aquatic environment.

Viruses also survive for extended periods in soil systems. Russian researchers have reported enterovirus survival for up to six months in soils. A comparison of the survival of bacteriophage f2 to that of poliovirus type 1 in sand under dry conditions, shows a longer persistence of the bacterial phage (Table 12.3).

More data are needed on virus indicators in soils and sediments. The lack of data has been probably due to our inability to detect viruses in these environments.

TABLE 12.3.   Survival of Poliovirus Type 1 (LSc) and Coliphage f2 in Dry Sand at Room Temperature[a]

Time (days)	% Survival	
	Poliovirus 1	f2 Phage
0	100.0	100.0
21	4.0	32.7
77	0.02	5.5

[a] Adapted from E. Lefler and Y. Kott (1974), *in: Virus Survival in Water and Wastewater Systems*, J. F. Malina, Jr. and B. P. Sagik, Eds., The University of Texas, Austin.

12.4 INDICATORS OF VIRUSES IN FOODS

As seen in chapter 11, little attention has been paid to food as a vehicle of viral disease. Infectious hepatitis and probably gastroenteritis are the only viral diseases that are known to be foodborne. The possibility exists that other enteric viruses may be transmitted via the food route. Part of the confusion stems from our inability to recover viruses efficiently from foodstuff. Although viruses have been recovered from shellfish, milk, meat, eggs, and vegetables, one hopes to be able to develop more sensitive techniques to obtain a more accurate picture of viral contamination of food. Animals may serve as a vehicle for viral transmission, since they can be infected by viruses of human origin (poliovirus, reovirus, coxsackievirus). Since viruses may potentially be present in our food supply, one is obviously concerned by their fate following food processing, mainly heat treatment and irradiation. It is also desirable to find a microbial indicator for the presence of enteric viruses in processed food.

Cliver, of the University of Wisconsin, proposed two criteria for the validity of an indicator in heat-processed food.

- *Criterion No. 1:* "The candidate indicator should always enter the food when the virus does and should never derive from sources irrelevant to viral contamination."
- *Criterion No. 2:* "During any subsequent heat treatment of the food, the candidate indicator should be killed or inactivated at the same rate or to the same extent as the virus."

A search for suitable indicators has shown that *E. coli* satisfies criterion No. 1, but does not comply with criterion No. 2. We have shown in Chapter 11 that thermal inactivation of microbes depends on many factors, including the type of microorganism, pH, salt concentration, and the type of food. For example, with respect to the latter, viruses are more resistant than enterococci in ice cream and butter and a reverse situation is observed in milk. Therefore, at the present time there is yet no reliable indicator of virus presence in heat-processed food.

Concern has been also raised over the presence of viruses in irradiated food (Chapter 11). The process of radappertization, which includes a thermal phase, should result in a significant destruction of viruses. A potential candidate indicator of virus presence in food treated by radappertization should comply with the two criteria outlined for heat-processed food.

Since viruses are relatively resistant to ionizing radiations, they are possibly present in food treated by radurization or radicidation. Indicators are badly needed to assess the safety of food treated by these processes.

In Chapter 11 we have stressed the role played by shellfish (oysters, clams, mussels) in the transmission of bacterial and viral diseases, particularly infectious hepatitis. Disease transmission is facilitated by three distinct features of shellfish, namely their ecology (they live in estuarine environment, often subject to pollution by human wastes), feeding habits (they have the ability to pump large quantities of water and thus concentrate viruses), and their preparation for consumption (they are often eaten raw or insufficiently cooked). Therefore it is of utmost importance to monitor shellfish beds for the presence of enteric pathogens. Since viruses and other pathogens are concentrated within the shellfish through the pumping of large quantities of contaminated water and since these pathogens survive longer within the mollusks than in the overlying water, it appears best to monitor the shellfish for the presence of the pathogens. Following the development of adequate detection techniques, virus presence has been demonstrated in shellfish meat and in shellfish-growing waters. In California mussels were found to harbor 25 to 1475 PFU/kg meat in areas which have been polluted by primary and secondarily treated sewage effluents. In a major European city 10% of marketed samples were found to contain enteroviruses. Two out of 17 samples of Texas oysters harbored viruses, namely, poliovirus 1 and echovirus 4. Poliovirus 3 was demonstrated in 1 out of 24 samples of Louisiana oysters. In France coxsackievirus A16 was detected in 10% of oyster and 20% of mussel samples. The adequacy of bacteriological standards has been investigated regarding the presence or absence of enteric viruses. On many occasions, it has been reported that coliform standards cannot be relied upon to assess viral pollution of shellfish and shellfish growing waters. This lack of correlation between virus and bacterial numbers is due to the higher survival of viruses in estuarine waters and in shellfish tissues. Enteric viruses may survive 3 to 6 times longer in mussel tissue than do coliform bacteria. Thus viruses have been frequently isolated from areas which were approved for shellfish harvesting according to bacteriological standards. Table 12.4 shows that viruses were always isolated from oysters considered safe according to the

TABLE 12.4.   Adequacy of Bacteriological Standards for Assessing the Virus Status in Oysters[a]

Station	No. of Samples	Fecal coliforms (MPN/100 g oysters)	% Positive Samples for Viruses
1	10	25	20
2	8	78	25
3	30	52	7
4	13	130	23

[a] Adapted from T. G. Metcalf (1975), in *Discharge of Sewage from Sea Outfalls*, A. L. H. Gameson, Ed., Pergamon Press, Oxford.

TABLE 12.5.   Adequacy of the Coliform Index for Assessing the Sanitary Quality (Viral Pollution) of Overlying Water in Shellfish Areas (Great South Bay, Long Island)[a]

Date	Total Coliforms (MPN/100 ml)	Fecal Coliforms (MPN/100 ml)	Number of Viruses (PFU/gal)
Open Area			
July 1976	4	4	8.0 (Polio 2, Echo 22, and Echo 11)
Aug. 1976	460	4	.12
Apr. 1977	150	15	2.9
June 1977	93	<3	No isolate
Closed Area			
July 1976	430	75	4.0
Aug. 1976	110	23	No isolate
Apr. 1977	2400	460	No isolate
June 1977	23	4	1.1 (Polio 1)

[a] Adapted from J. M. Vaughn and E. F. Landry (1977), "An assessment of the occurrence of human viruses in Long Island aquatic systems." Report No. BNL 50787, UC-11, Brookhaven National Laboratories, Upton, N.Y.

fecal coliform index that allows no more than 230 fecal cloiforms per 100 g of meat.

In Long Island, viruses have been detected in waters open to shellfish harvesting (Table 12.5), and thus judged to be acceptable by the coliform standard. Therefore, it appears that coliform standards are unreliable with regard to viral pollution of shellfish and overlying waters.

Some have considered the use of coliphages as indicators of enteric viruses in shellfish and in estuarine waters. A study of coliphage occurrence in the Great Bay estuarine system has essentially shown that one cannot rely on coliphage index to assess viral pollution of water, sediments, and shellfish. Field sampling has revealed that coliphage were isolated while enteric viruses could not be detected and vice versa. Although coliphages display a survival pattern similar to that of enteroviruses, their replication in the estuarine environment (Figure 12.2) is possible. This shortcoming would make the coliphage index unsuitable for indicating the viral quality of water and shellfish.

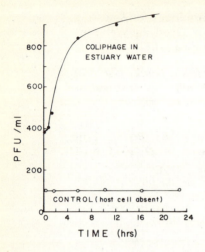

FIG. 12.2.   Replication of coliphage in estuarine water (dialysis bags containing known volumes of sterilized seawater were inoculated with *E. coli B* and 100 coliphage/ml. They were incubated in estuarine water at 18°C. Control bags were not seeded with host cells) Adapted from J. M. Vaughn and T. G. Metcalf (1975), *Water Res.* **9:**613–616.

12.5   PROPOSED STANDARDS OR GUIDELINES FOR VIRUSES

Water quality criteria (i.e., the relationship between the level of a given indicator and potential health hazards) enable us to set *guidelines* and *standards* for pathogens. A *water quality guideline* is "a suggested upper limit for the density of a given indicator which is associated with unacceptable health risks." A *water quality standard* is defined as "a guideline that has been fixed by law." In the United States there are yet no standards pertaining to viruses in water, soils, sludges, and food; there are only suggested guidelines.

12.5.1   Basis for Establishing Water Quality Regulations

Regulations concerning water quality (chemical, physical, and microbiological parameters) are generally based on the following:

- *Background Levels.* However, little is known about the potential health aspects of the pollutant under consideration. Furthermore, the background concentration may vary with time and space.

- *Aesthetics.* This is the oldest way of judging water quality prior to the development of sophisticated quantitative tools. For example, taste and odor were judged on that basis.

- *Detection Limit.* This approach may be expensive and presently only within the reach of large water treatment plants. Virus guidelines are based on this approach. We have now the technical capability of detecting as low as 1 plaque-forming unit (PFU) in 100 gallons (378 liters) of drinking water. (This topic has been examined in Chapter 5). However, this approach does not involve any relationship to health hazards.

- *Treatment Technology.* A water quality standard may be based on the performance of wastewater and water treatment plants regarding reduction of microbial pathogens. This system does not also take into account the human health aspect. Present bacterial standards are based on treatment technology.

- *Relationship with Health Effects.* A standard based on health effect must rely on toxicological (e.g., chemical pollutant) and epidemiological (e.g., microbial pollutant) data. Establishing this relationship is a rather complex, expensive and time-consuming undertaking. Epidemiological methods are not sensitive enough to detect low-level transmission of enteric viruses and supporting data are lacking on this subject.

12.5.2   Water Quality Standards and Guidelines

We will examine five categories of water with regard to bacterial and, if available, viral standards or proposed guidelines. These categories are drinking water, recreational waters, shellfish growing waters, irrigation water, and reclaimed water.

Drinking Water. The quality of drinking water is based on the coliform standard of <1 coliform/100 ml (arithmetic mean of all samples taken during 1 month). Although this standard was instrumental in the decline of bacterial diseases (e.g., cholera, typhoid fever), it is of uncertain role regarding viral disease occurrence. The need to monitor viruses derives from the fact that bacteria are less resistant than viruses to water treatment processes, particularly disinfection. There are presently no standards regulating the levels of viruses in drinking water. A recent law (PL-93-523) passed in 1974 calls for national interim drinking water standards, but no reference was made to viruses. A reasonable virus standard is one which can be met by present water treatment plants and verified by our detection techniques. Based on detection limits a guideline for viruses was suggested and it calls for 1 or less than 1 PFU/100 gal (378 liters) or drinking water. European and World Health Organization "standards" call for 0 PFU/10 liters.

Recreational Waters. Recreational waters are often polluted by sewage treatment plants and septic tank effluents, and by stormwater runoff. The Federal Water Pollution Control Administration (FWPCA) established guidelines based on levels of total coliforms (<1000/100 ml) and fecal coliforms (<200/100 ml). These guidelines have been criticized on the basis that they are not supported by solid epidemiological data. The establishment of microbial standards for bathing beaches is a matter of controversy. This is due to numerous methodological obstacles in epidemiological surveys. With respect to viruses, a guideline of 1 PFU/10 gal. (37.7 liters) of recreational water has been proposed but there are no epidemiological data to support it. It was recently demonstrated that swimmers in polluted marine waters had higher rates of gastrointestinal diseases than nonswimmers. Rotaviruses may be responsible for some of the enteric disturbances.

For swimming pool waters, the free residual chlorine is the most important indicator of water quality. Guidelines based on levels of bacterial indicators (total coliforms and standard plate count) have also been proposed. These guidelines call for 0 coliform/100 ml and a standard plate count of no more than 100/100 ml of swimming-pool water. No virus guidelines or standards appear necessary for this category of water and maintenance of a sufficient free chlorine residual should be emphasized.

Shellfish-growing Waters. The sanitary quality of shellfish-growing waters is based on bacterial standards. For approved waters the coliform MPN (most probable number) should be less than 70/100 ml with no more than 10% of the samples exceeding 230/100 ml. Shellfish meat should contain no more than 230 fecal coliforms/100 g. There are presently no guidelines for viruses in shellfish or in shellfish growing waters. These guidelines appear necessary since bacterial indicators are unreliable with respect to viral pollution.

Irrigation Water. The use of sewage effluents for spray irrigation of crops may potentially result in the production of pathogenic aerosols and contamination of crops, particularly those eaten raw (e.g., lettuce, radishes).

TABLE 12.6. Conformation of Drinking Water Samples (Conventional vs Reclaimed Water) to Proposed Microbiological Quality Standard[a]

		% Samples Failing to Conform to Proposed Limits	
Parameter	Proposed Standard	Conventional Supplies	Reclaimed Water
Standard plate count	100/ml	51.0	37.3
Total coliforms	0/100 ml	23.7	14.3
Fecal coliforms	0/100 ml	13.2	8.3
Virus count	0/10 liter	0.3	0.0

[a] Adapted from W. O. K. Grabow (1977), *in: Bacterial Indicators/Health Hazards Associated with Water,* A. W. Hoadley and B. J. Dutka, Eds., ASTM (STP 635), Philadelphia, Pa.

TABLE 12.7.   Microbiological Standards or Guidelines for Water

Category of Water	Bacterial Standards or Guidelines[a]	Proposed Guidelines for Viruses	Comments on Viruses
Drinking water	<1 TC/100 ml	<1 PFU/100 gal	Suggested by J. L. Melnick
		0 PFU/10 liters	WHO international and European standards
Recreational water	<1000 TC/100 ml	<1 PFU/10 gal	Suggested by J. L. Melnick[b]
	<200 FC/100 ml (Established by FWPCA)		
Swimming pool water	0 TC/100 ml <100 TPC/100 ml	No guideline	Free chlorine residual is the most important water quality indicator
Shellfish-growing water	<70 TC/100 ml (water) <230 FC/100 g (shellfish meat)	No guideline	Need efficient techniques for recovery of viruses from shellfish meat
Reclaimed water	TPC = 100/ml TC. = 0/100 ml FC. = 0/100 ml	0/10 liters	Suggested by W. O. K. Grabow for South Africa
		0/100–1000 liters	WHO, 1978

[a] TC = Total coliform; FC = Fecal coliform; TPC = Total Plate Count
[b] Guideline adopted by the State of Arizona for full body contact (swimming, water skiing …). For partial body contact (fishing, boating …) the set limit is 125 PFU/40 liters.

Microbial aerosols are not regulated by any bacterial or viral standard or guideline. Viruses have been shown to be hardier than bacterial indicators (total and fecal coliforms), and their sampling is costly and complicated. This is the reason why models have been proposed to predict their concentration in air following spray irrigation with wastewater effluents (see Chapter 9).

Some guidelines have been suggested for regulating the sanitary quality of spray irrigation water. These guidelines, when available, vary from one state to another. The state of California guidelines specify that no more than 23 coliforms/100 ml should be allowed in spray irrigation water for crops that are processed (to kill pathogens) prior to consumption, and no more than 2.2 coliforms/100 ml regarding those crops that are eaten raw. A World Health Organization standard specifies that no more than 100 coliforms/100 ml should be allowed in 80% of the water samples.

Reclaimed Wastewater. Work conducted in South Africa on wastewater reclamation systems (Windhoek and Stander plants) has demonstrated that it is now technologically possible to produce microbiologically safe reclaimed water. It has been advanced that the reclaimed water is of better quality than drinking water produced by conventional methods (Table 12.6). Epidemiological investigations have revealed that the reclaimed water was never associated with bacterial (typhoid fever, shigellosis) or viral (infectious hepatitis, non-bacterial gastroenteritis) diseases. In wastewater reclamation plants, the lime treatment process (pH > 11.0) is very efficient against viruses. It was found that the total plate count was the most suitable indicator for assessing the quality of reclaimed water. The following microbial standards were suggested: total plate count, 100/ml; total coliforms, 0/100 ml; fecal coliforms, 0/100 ml; coliphage, 0/10 ml; and enteroviruses, 0/10 liters.

A summary of bacterial and viral standards or regulations for various categories of water is shown in Table 12.7.

In conclusion, the setting of virus standards necessitates the development of efficient, inexpensive, and rapid techniques for virus detection in water or any other environmental samples (soils, sediments, sludges). For economic reasons the water samples could be concentrated on site, and the concentrates could be shipped, under appropriate conditions, to a central virus detection laboratory. Virus standards may also be based on a relationship between virus levels and human health. This relationship is difficult to establish at the present time.

12.6 CONCLUSIONS

We have covered all the possible indicators that may be used for assessment of viral pollution. We realize by now that there is no ideal indicator except the virus itself. Some of the indicators are either more sensitive or more resistant (e.g., bacterial spores) than viruses to environmental stresses and disinfection. This may lead to a false sense of security or to an exaggeration of the virus threat. Once a suitable indicator has been found, one may eventually have to correlate its density in water to disease occurrence within the community.

We may also choose another direction, that is, the direct enumeration of viruses in environmental samples. This approach is, however, unpractical at the present time, due to the numerous complications encountered in virus detection technology. In the end we will probably give up our hope of finding a universal indicator for viruses and resign ourselves to the use of different indicators for different purposes. For example, yeast and mycobacteria may serve well as indicators for viruses during the chlorination process. Phages could eventually be suitable as indicators in wastewater treatment plants.

It has been suggested that the vaccine strain of poliovirus may well serve as an indicator for other enteroviruses and possibly for the entire enteric virus group. This particular strain is shed in large numbers by vaccinated persons, is relatively stable to environmental stresses, is relatively safe to handle and may be easily detected in environmental samples. Although poliovirus has not been always detected in sewage samples despite routine administration of live attenuated poliovirus vaccine, it has been selectively isolated on a human amnion cell line containing antisera against other viruses. However, does poliovirus behave similarly to hepatitis A virus under environmental stresses? This problem remains a real challenge to environmental virologists.

12.7 FURTHER READING

Berg, G. Ed. 1978. *Indicators of Viruses in Water and Food.* Ann Arbor Science Publications, Ann Arbor, Mich.

Cabelli, V. 1978. New standards for enteric bacteria. In: R. Mitchell, Ed. *Water Pollution Microbiology,* Vol. 2. Wiley, New York.

Gameson, A. L. H., Ed. 1975. *Discharge of Sewage from Sea Outfalls.* Pergamon, Oxford.

Geldreich, E. E., and N. A. Clarke. 1973. The coliform test: A criterion for the viral safety of water. In: *Virus and Water Quality: Occurrence and Control,* V. L. Shoeyink et al., Eds. 13th Water Qual. Conf., University of Illinois, Urbana–Champaign, Ill.

Grabow, W. O. K., B. W. Bateman, and J. S. Burger, 1978. Microbiological quality indicators for routine monitoring of wastewater reclamation plants. *Prog. Water Technol.* **10**:317–327.

Hoadley, A. W., B. J. Dutka, Eds. 1977. *Bacterial Indicators/Health Hazards Associated with Water.* American Society for Testing and Materials, Philadelphia, Pa.

Metcalf, T. G. 1978. Indicators for viruses in natural waters. In: *Water Pollution Microbiology,* Vol. 2. R. Mitchell, Ed. Wiley, New York.

Shuval, H. I. 1976. Water needs and usage: The increasing burden of enteroviruses on water quality. In: G. Berg et al., Eds. *Viruses in Water*, American Public Health Association, Washington, D.C.

Sproul, O. J. 1976. Standards for viruses in effluents, sludges and ground and surface waters. In: L. B. Baldwin, J. M. Davidson, and J. F. Gerber, Eds. *Virus Aspects of Applying Municipal Waste to Land*, University of Florida, Gainesville, Fla.

Index